KT-583-230

Analytical Physiology of Cells and Developing Organisms

Dedicated to

PROFESSOR C. H. WADDINGTON, F.R.S.

Ah, but a man's reach should exceed his grasp,
or what's a heaven for?

"Andrea del Sarto"
R. Browning

Analytical Physiology of Cells and Developing Organisms

B. C. GOODWIN

School of Biological Sciences, University of Sussex

1976

ACADEMIC PRESS INC. (LONDON) LTD.
24/28 Oval Road,
London NW1

United States Edition published by
ACADEMIC PRESS INC.
111 Fifth Avenue
New York, New York 10003

Library of Congress Catalog Card Number 76–1081
ISBN: 0–12–289360–3

PRINTED IN GREAT BRITAIN
J. W. Arrowsmith Ltd., Bristol

PREFACE

My intention in writing this book has been to construct a picture of the physiological behaviour of cells and developing organisms based upon the insights of molecular biology on the one hand and mathematical model-building on the other. The use of the term analytical physiology derives from this two-fold basis, analysis being understood in both its experimental and its mathematical senses, with complementary interaction between them. The first three chapters of the book are dominated by the view of cell behaviour in terms of the control metaphor, wherein homeostasis and adaptation are regarded primarily as manifestations of positive and negative feedback processes, and the stability of the living system is interpreted in informational rather than in energetic terms. This view emerged during the 1950s and developed into a golden age of control theory and cybernetics in biology, but it is now undergoing a rapid transformation into something else, not yet named. This may be the flowering of an indigenously biological metaphor which is dominated by the concepts of structure, form and transformation rather than of purely dynamical stability. This would seem to be an appropriate development in biology, a subject in which morphology has always been a primary feature. The tendency to denigrate comparative anatomy and descriptive embryology during the past few decades is consistent with an analytical approach to the organization of living systems, and was a necessary part of the development of molecular biology. The success of this approach is extremely impressive, and constitutes one of the most exciting developments in the history of biology.

However, the picture of the organism which we get from molecular biology and the control metaphor has its limitations, and these become progressively more evident as the book proceeds. One can get away with very little reference to structure and morphology as long as one is concerned with the purely homeostatic and adaptive properties of cells, although even here it soon becomes apparent that size and shape must be taken into account, if only minimally. When one comes to study the stability of tissues and the phenomena of regeneration and morphogenesis, however, form becomes a primary aspect of the problem and the dynamical aspects of physiology become extended and organized in space. From Chapter 4 onwards, the consideration of such spatio-temporal relationships becomes progressively prominent. The dynamical models of developmental processes considered in Chapters 5 and 6 still derive from molecular biological postulates, but diffusion and wave propagation begin to emerge as important space-ordering processes. The study of morphogenetic mechanisms is one which is now developing very rapidly, and we shall certainly witness a dramatic transformation of viewpoint in the next few years from the relatively static space-ordering models which still tend to dominate experimental study in this field, to dynamic and flexible space–time processes, some of whose properties I attempt to anticipate in these chapters.

In an effort to unify the ideas developed throughout the book and to provide a context within which they can be transcended and linked with other areas of study which I think have natural affinities with developmental biology, I explore in the last chapter an approach to the study of organismic behaviour in which I attempt to combine a phenomenological attitude with an analytical one, rather in the spirit of contemporary structuralism. Here the organism is regarded as a cognitive system, adapting and evolving on the basis of knowledge about itself and its environment. The dynamical modes of an organism's behaviour are represented as manifestations of co-operative or collective activity among cognitive units, development being seen as the orderly unfolding of these modes within a structurally stable, knowledge-using system.

The general mood of the book is one which regards conceptual (including experimental) and mathematical analysis as continuous and mutually reinforcing, but separable according to the taste of the reader. Unlike my previous book, in which I was preoccupied with the search for a formal treatment of physiological processes within cells, the primary material in this volume is biological, with a variety of mathematical devices employed to analyse in more detail certain aspects of the behaviour of cells and developing systems. Having started my career as an experimental biologist involved in physiological research before being diverted by the formal beauty of mathematics and the elegance of statistical thermodynamics, I feel that the measure in which this book is dominated by experimental problems is the measure of my return to original preoccupations.

As to structure, the organization of the material is hierarchical, as befits a biological analysis. A brief description of pre-control view of the stability properties of biochemical networks is followed by discussions of metabolic regulation, macromolecular regulation, the cell cycle, biological clocks, cell population control and morphogenesis in unicellular and multicellular organisms. Much of this material comes from various courses given to generations of students at the University of Sussex, to whom I am grateful for reactions, corrections and clarifications. The treatment of the material is in no sense comprehensive, the selection being a personal and therefore idiosyncratic one. I attempt to make acknowledgements wherever appropriate, but much of my indebtedness to individuals is of an educational nature. However, there is one outstanding personal debt of gratitude which I owe, and this is to the late Professor C. H. Waddington, to whom the book is dedicated. It was his belief in the importance of theoretical analysis and mathematical model-building in biology which first provided me with the opportunity of exploring beyond the conceptual horizons which I had experienced in experimental research. This same belief of Waddington's gave rise to the series of conferences at the Villa Serbelloni on Lake Como which constituted a unique educational experience in interdisciplinary discussion and has resulted in the volumes entitled 'Towards a Theoretical Biology'. The 'Towards' in this title was symptomatic of Waddington's thought, which was always reaching beyond accepted ideas and involved a constant flirtation with the heretical. However, this play with heterodox fire was firmly based upon a deep familiarity with the detailed behaviour of organisms. It is this type of balance between knowledge and vision after which I aspire, and whatever quality this book may have in this respect is due in no small measure to inspiration received directly and indirectly from C. H. Waddington. Another acknowledgement I would like to make is to the Science Research Council of Great Britain, whose generosity in financing an Analytical Biology Research Group at Sussex made possible the appointment of individuals with mixed scientific parentage and creative insight to visiting and research positions in the area of biological analysis of the type described in this book. To my friends and colleagues at the University I am indebted as well, but particularly to Drs John Dowman, Gerald Webster and Keith Oatley.

The last chapter of the book derives from ideas explored during a sabbatical year at the National University of Mexico. The collaborative work carried out there with physicists and biologists was of the greatest value to me, and my thanks for many discussions go to Drs Manuel Berrondo, Germinal Cocho, Octavio Novaro, Rafael Perez-Pascual, Ruy Perez-Tamayo and Gustavo Viniegra-González. To the heads of the Departments of Theoretical Physics and of Biomedical Research, Drs M. Moshinsky and J. Mora, I am indebted for hospitality and assistance; while to the Royal Society I owe a debt of gratitude for a Leverhulme Visiting Prefessorship which made the visit to Mexico a possibility. And lastly, although they are unlikely to be in the least interested, I am grateful for the serene magnificence and beauty of Istaccihuatl and Popocatapetl whenever they were in view from my window in the Torre de Ciencias, for they were a potent source of detached inspiration.

September, 1976

B. C. GOODWIN
Sussex University

CONTENTS

The Garden of Forking Paths is an incomplete, but not false, image of the universe as Ts'ui Pen conceived it. In Contrast to Newton and Schopenhauer, your ancestor did not believe in a uniform, absolute time. He believed in an infinite series of times, in a growing, dizzying net of divergent, convergent and parallel times. This network of times which approached one another, forked, broke off, or were unaware of one another for centuries, embraces all possibilities of time. We do not exist in the majority of these times; in some you exist, and not I; in others I, and not you; in others, both of us. In the present one, which a favourable fate has granted me, you have arrived at my house; in another, while crossing the garden, you found me dead; in still another, I utter these same words, but I am a mistake, a ghost.

"Labyrinths"
Jorge Luis Borges

Chapter 1

STABILITY AND REGULATION IN THE METABOLIC SYSTEM

THE METABOLIC NETWORK

OUR story begins in the years immediately before control theory became dominant, that is, before the early 1950s, when the most obtrusive feature of cellular metabolism was its sheer complexity. Thirty years of biochemistry had been necessary before some coherent picture began to emerge regarding the inter-related pathways whereby one metabolic species is converted into another in living organisms, each reaction being catalysed by a specific enzyme. It had become evident that there must be hundreds of different enzymes in each cell and it was conjectured that the exact metabolic state of any cell was determined by the types and the amounts of each of these organic catalysts, together with the concentrations of precursors or nutrients available to the cell. A useful, but somewhat inaccurate representation of this picture of branching and inter-connected metabolic pathways is given in

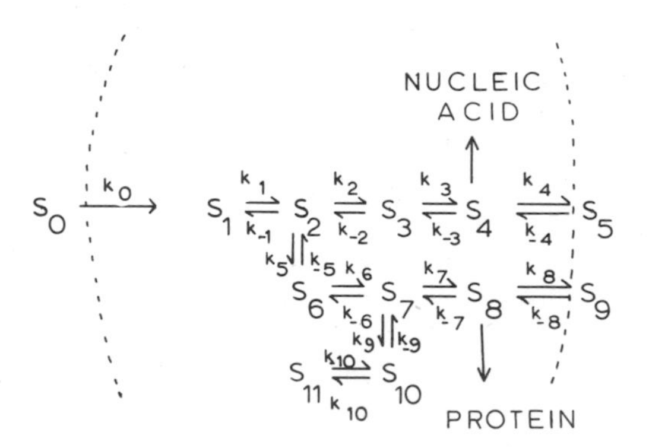

Figure 1.1. A linear picture of the branching pathways characteristic of the metabolic system of a cell. The dotted lines represent the membrane boundaries, S_0, S_5, and S_9 being external nutrients. S_8 and S_4 are precursors for protein and nucleic acid synthesis, respectively.

Figure 1.1. This is simply a kinetic description of what one sees on a metabolic pathway chart, with some simplifications that I will consider in a moment.

One can imagine such a network expanded to accommodate several hundred different metabolic species, S_i, with associated rate constants for forward and backward reactions. These rate constants include enzyme concentrations, so that the first assumption made in representing a metabolic network in this form is that these concentrations are not changing. In looking for some analytical insight into the collective behaviour of the variables of intermediary metabolism, the first question we encounter, then, is: for what period of time is it legitimate to assume that enzyme concentrations in the cells will remain constant? The phenomenon of enzyme adaptation, wherein a culture of micro-organisms adapts to a change of carbon source (galactose instead of glucose, say) by the appearance of a new enzyme activity has been known for nearly a century and tells us that enzyme activities in cells do change, but this process requires between 20 minutes and a few hours in organisms such as bacteria or yeast. Thus we can put a very rough upper bound to the time interval over which enzyme activities may be assumed to be constant in such organisms at a few minutes. Since we are interested in variations in S_1, the metabolites, we must choose a lower bound which gives enough time for change in these variables. This we can estimate from enzyme turnover numbers (the average number of molecules of substrates an enzyme molecule can convert into product in one second) which lie in the range 10^2–10^5, with a usual figure about 10^3. For a measurable change in the concentration of some metabolite such as glucose 1,6-diphosphate or aspartate, then, a minimum period of at least a few seconds would be required. So in considering the behaviour of a metabolic network of the type represented in Figure 1.1, we are restricting our attention to processes in the time range of seconds to minutes. This corresponds to the experimental periods used by enzymologists and those studying the processes of intermediary metabolism.

Now we can be somewhat more analytical in our procedure, using little more than the information already introduced. We need to know first what the expected concentration of a metabolite is likely to be in a cell, and this we can deduce simply from the Michaelis constants of enzymes. This is because these constants, telling us the concentration of substrates that give half-maximal velocities of the enzyme-catalysed reactions, may be expected to correspond to the effective metabolite level in the living system where the enzyme is operating. Of course this can only be an estimate, since the properties of a purified enzyme *in vitro* are bound to be different from those of the *in vivo* catalyst, but estimates are all we want at the moment. A typical Michaelis constant is about 10^{-3} M, which we then take to be a mean value of a variable S_1. Next we need an estimate of enzyme concentrations in cells. This, of course, varies widely between different enzymes, but an average value is about 10^{-8} M. Using a mean enzyme turnover number of 10^3

molecules/s, we find that the rate of change of substrate produced by the enzyme is $10^{-8}\ \text{M} \times 10^{3}\ \text{s}^{-1} = 10^{-5}\ \text{M s}^{-1}$. Consider now the equation for one of the steps in the reaction sequence of Figure 1.1, say that for the variable S_1:

$$\frac{dS_1}{dt} = k_0 S_0 + k_{-1} S_2 - k_1 S_1. \tag{1.1}$$

We take S_0 to be a constant, representing some nutrient source term (e.g., glucose). Let us take S_2 also to be a constant, an approximation which would be valid if we were perturbing the system slightly from its study state by changing S_1 (say by injection), then following the initial stages of the return to the steady state. In this case we can solve the equation for S_1. Writing $a = k_0 S_0 + k_{-1} S_2$, constant, we get

$$\frac{dS_1}{dt} = a - k_1 S_1, \tag{1.2}$$

whose solution is

$$S_1 = \frac{a}{k_1} + b\, e^{-k_1 t} \tag{1.3}$$

where b is determined by the initial conditions. If the experiment consists in a perturbation of S_1 which causes a transient increase over its steady state value which is a/k_1, the size of this increase is b. Using our previously estimated values of $S_1 = 10^{-3}\ \text{M}$ and the estimated value for the rate of change of substrate, $k_1 S_1 = 10^{-5}\ \text{M s}^{-1}$, we find that $k_1 = 10^{-2}\ \text{M s}^{-1}$. We may now ask the more precise question: how long will it take for the initial perturbation to have decreased to e^{-1} of its original value? Evidently this time is given by the value of t for which $k_1 t = 1$, i.e., $t = 1/k_1 = 10^2$ s, or about two minutes. This is known as the relaxation time of the system described by equation (1.2). Thus we get a more detailed answer to the question: what is the characteristic time for metabolic experiments? And we may then use this time to define what I will refer to as the metabolic system: that set of interconnected variables undergoing enzyme catalysed or more generally protein-mediated transformations which change significantly in time periods too short for significant variations in macro-molecular concentrations. In the study of a system as complex as the living cell, it is essential to use analytical devices of this kind to dissect the system temporally if we are to get any dynamical insight into its total operation. The traditional biological disciplines have in fact identified themselves by precisely these criteria, reflecting accurately what we presume to be the organizational principles of the living system itself. What has been identified above is the temporal aspect of the hierarchical organization which has always been recognized as the foundation of biological order. We will see how this dynamical hierarchy unfolds in the course of our analysis, providing insight into adaptive and developmental behaviour.

Stability and Mass Action in the Metabolic System

Equation (1.3) shows us that the steady state of the kinetic system described by equation (1.2) is stable to perturbation, because of the negative exponential term. Graphically, the response is as shown in Figure 1.2. This behaviour arises from a very basic assumption about kinetic reactions, embodied in the law of mass action, and giving rise to the term $-k_1S_1$ in equation (1.2): the rate of a reaction is proportional to the concentration of

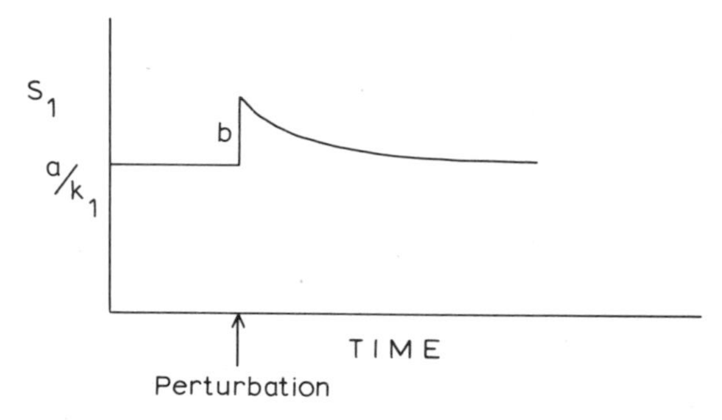

Figure 1.2. The response of a metabolic intermediate, S_1, to a small perturbation of amplitude *b*.

reactants. Thus the more S_1 there is, the faster is it converted to its product. This is true of all steps in the reaction sequence of Figure 1.1 and so one is led to the conjecture that a network of this kind is probably stable to the perturbation of any one of its variables, or indeed to any set of them. This was in fact shown to be true by Hearon (1952), and for precisely the reason conjectured: the law of mass action. However, we must now consider the consequences of a second assumption that has been used in representing intermediary metabolism in the form shown in Figure 1.1, which is that each reaction is regarded as first order in the substrates. We know that many reactions are bimolecular: aspartate and carbamyl phosphate are both involved in the reaction catalysed by aspartate transcarbamylase to produce carbamyl aspartate, for example. It is possible that *in vivo* such reactions are quasi-linear, one substrate being present in virtually saturating concentrations, but it seems unlikely. Therefore the equations we use to represent the kinetics, such as (1.1), will in general include products of concentrations of two or more variables instead of being linear. Hearon's result was for linear systems for which there is only a single steady-state solution. For the metabolic system generally, with quadratic and higher order terms in the kinetic equations, we cannot draw firm conclusions because it is simply not possible to obtain general solutions and to examine the stability of the different steady states. One encounters immediately severe mathematical difficulties with such non-linear systems. There are, however, some considerations of a general analytic nature which suggest that non-linear networks will still have locally stable steady states. One such is provided by Newman (1971). The older literature on this problem is also of interest (Kacser, 1957; Denbigh *et al.*, 1948), but it is inconclusive as regards the general case. A recent study of the qualitative types of instability which second and third order chemical systems can show is given in an interesting paper by Tyson and Light (1973), in which they demonstrate that only very limited forms of behaviour are available to such systems other than the familiar asymptotically stable states of the type we have been considering, exemplified in the response shown in Figure 1.2. One may say that despite some significant advances the problem of analysing complex non-linear systems remains nearly as intractable now as it did in the early 1950s, and it is fortunate that insight into the organization of metabolic processes in cells was not dependent upon theoretical analysis but upon experiment. In the early 1950s some observations were made on the behaviour of metabolic mutants in bacteria which transformed the picture of Figure 1.1 into a totally new form, ushering in the cybernetic age of cellular physiology.

Feedback Inhibition in Metabolic Pathways

Let us consider what would be the expected behaviour of a system of the type shown in Figure 1.1 in the absence and in the presence of an exogenous source of some metabolite such as S_9. Suppose it to be an amino acid, and our system to be a bacterium, say *Escherichia coli*. If the amino acid is present in the growth medium it will be transported across the bacterial membrane and will enter the internal pool of the amino acid designated S_8, there being available for activation and incorporation into protein. Since metabolic reactions are thermodynamically reversible (the rate constants for forward and backward reactions are non-zero), the amino acid can undergo transformation to S_7, S_6, etc. The only effect of S_9 being present is a mass action effect in the system. All variables will continue to be present within the system, but at an increased steady state value over that when S_9 is absent.

With the discovery of sexuality in bacteria by Lederberg in 1946, genetic techniques of analysis became available for the study of bacterial metabolism and a period of intense and immensely fruitful investigation was begun.

By 1955 it had been observed that mutant strains of bacteria with a block late in a biosynthetic sequence due to the presence of a defective enzyme tended to accumulate metabolic intermediates prior to the block (Roberts *et al.*, 1955). For example, if the mutant enzyme was that catalysing the reaction

$$S_7 \underset{k_{-7}}{\overset{k_7}{\rightleftharpoons}} S_8$$

so that $k_7 = k_{-7} = 0$, then precursors S_6 and S_7 were observed to be present in relatively large amounts. The bacteria would not, of course, be able to grow in the absence of exogenous amino acids. This observation is perfectly consistent with our picture as far as it goes, suggesting only that the reaction from S_2 to S_6 is an exergonic one (liberating free energy) so that the equilibrium is in favour of an accumulation of the intermediates leading to S_8, the end product of the sequence. However, it was then established that if the end product is provided exogenously the precursors disappear from the bacteria, which were then capable of growing (Novick and Szilard, 1954). This is not consistent with our mass action picture, since the presence or absence of S_8 in the cell should make no difference to S_6 and S_7 when $k_7 = k_{-7} = 0$. It strongly suggested that the end product was having a specific effect on the enzyme catalysing the first step in the reaction pathway, from S_2 to S_6.

Direct confirmation of this followed in 1956 when Umbarger showed that threonine dehydrase, the first enzyme of five in the pathway from threonine to isoleucine, the end product of the sequence, was strongly and specifically inhibited by L-isoleucine. He called this, appropriately, end product inhibition. In the same year, Yates and Pardee (1956) reported the specific inhibition of aspartate transcarbamylase, the first enzyme in the pathway from aspartate to the pyrimidines, by cytidine triphosphate. It is of interest to observe that the forward reactions in both these instances are strongly exergonic, making them physiologically irreversible (i.e., essentially irreversible under physiological time scales and conditions). Exit of end product from the pathway is either by incorporation into macromolecules (protein or nucleic acid for the respective enzymes), or by degradation, thus producing a one-way flux along the biosynthetic sequence. It is this unidirectional property which makes control of flow rate by a single enzyme a possibility in these cases. The exergonic nature of the first step makes this the thermodynamically efficient place to exercise control. Metabolic logic leads to the same conclusion, since if the end product is available from an alternative source (e.g. from the nutrient outside the cell), then none of the metabolic intermediates is required along the pathway in so far as their only metabolic purpose is to provide end product. It is gratifying to find organisms behaving logically as well as efficiently according to our lights, suggesting that we have grasped at least one of their essential principles of organization. One could see what the consequence would be if this principle were not followed. Consider what would happen if, for example, lactate dehydrogenase were inhibited by lactate, product of pyruvate reduction by reaction with NADH. This reaction is physiologically reversible, allowing lactate to accumulate under certain circumstances, such as strenuous muscular activity, and then to be oxidized when conditions allow. Since there is no metabolic exit for lactate other than via pyruvate, lactate inhibition of LDH would create a pool of trapped product causing considerable physiological distress. Thus we see that physiologically reversible metabolic steps are there for good reasons and that reversibility is incompatible with control, which provides a throttle for a one-way process.

There are, of course, metabolic sequences along which flow can occur in either direction according to physiological demand. A classic example is glycolysis-gluconeogenesis. One might suppose that in such an instance, no control would be exercised and that mass action would be allowed to determine the net flux rate according simply to pool sizes of the metabolites at either end of the pathway. This would work in its own rather sluggish way, but clearly it would be very inefficient. This is because some of the steps in glycolysis, for example, are necessarily exergonic, such as the phosphorylation of fructose 6-phosphate by phosphofructokinase (using ATP) to give the product fructose-1,6-diphosphate (FDP). Reversal of this reaction by mass action alone would require the accumulation of very large amounts of FDP, which would disturb cellular pH and osmotic values as well as accumulating a great deal of energy in one component of the system. Therefore an alternative strategy is called for, which is to transform FDP in another way which avoids these disadvantages. Cells do so by exploiting the alternative pathway from FDP to F6P, which proceeds by hydrolysis and so is again an exergonic reaction. The enzyme which catalyses this transformation is phosphatase. Thus it appears that the price paid in the form of a free energy loss due to the hydrolysis of FDP releasing inorganic phosphate instead of ATP as would occur in the other reaction, is readily paid by the cell in order to achieve efficient reversal of this step in the sequence. The enzymes phosphofructokinase and phosphatase then become potential sites of metabolic control by processes such as end product inhibition. We will see later that this is precisely what one finds. Thus regulation can occur in a reversible sequence of metabolic steps, but it operates according to the same principles as those operating in an irreversible or one-way sequence such as the biosynthetic pathways involving threonine dehydrase and aspartate transcarbamylase, as discussed above.

Allosteric Behaviour of Enzymes

It was recognized immediately after the discovery of end product inhibition that an essentially new principle of enzyme behaviour must be involved. Inhibition of enzyme activity by competitive or non-competitive inhibitors had been known and studied for many years, and both were reasonably well understood. A classical case of competitive inhibition is the effect on succinate dehydrogenase activity of malonic acid, sterically similar to succinic acid as we see from Figure 1.3. The kinetic treatement of the interaction between an enzyme, *E*, and its substrate, *S*, and such a competitive

```
        O
        ||
        C - O⁻                    O
        |                         ||
H  -    C - H                     C - O⁻
        |                         |
H  -    C - H             H  -    C - H
        |                         |
        C - O⁻                    C - O⁻
        ||                        ||
        O                         O

  SUCCINIC ACID             MALONIC ACID
```

Figure 1.3. Succinic and malonic acids are sterically very similar and so compete for the active site on the enzyme succinic dehydrogenase.

inhibitor, *I*, is as follows. *E* combines with *S* to give the intermediate complex *ES*, which then undergoes transformation to $E+P$, enzyme plus product, according to the scheme

$$E+S \underset{k_{-1}}{\overset{k_1}{\rightleftharpoons}} ES \xrightarrow{k_2} E+P \tag{1.4}$$

where we are considering only the initial stages of the reaction before sufficient *P* has accumulated to make the back reaction to *ES* appreciable. Enzyme also combines with inhibitor to give an inactive form, according to the scheme

$$E+I \underset{k_{-3}}{\overset{k_3}{\rightleftharpoons}} EI. \tag{1.5}$$

From these reactions we may write the following kinetic equations:

$$\frac{\mathrm{d}[ES]}{\mathrm{d}t} = k_1[E][S]-(k_{-1}+k_2)[ES] \tag{1.6}$$

and

$$\frac{d[EI]}{dt} = k_3[E][I] - k_{-3}[EI]. \tag{1.7}$$

There is also a conservation condition on the enzyme, viz.,

$$[E]_0 = [E] + [ES] + [EI] \tag{1.8}$$

where $[E]_0$ is the total amount of enzyme initially added to the reaction. At the steady state of the reaction we may take

$$\frac{d[ES]}{dt} = \frac{d[EI]}{dt} = 0$$

and solve the three equations (1.6), (1.7) and (1.8) for $[ES]$, as a function of S and I. From (1.6) we get

$$[ES] = \frac{k_1}{k_{-1}+k_2}[E][S] = K_1[E][S] \qquad \text{where } K_1 = \frac{k_1}{k_{-1}+k_2},$$

while from (1.7) we find

$$[EI] = \frac{k_3[E][I]}{k_{-3}} = K_2[E][I], \qquad \text{where } K_2 = \frac{k_3}{k_{-3}}.$$

Using these in equation (1.8) we have the result

$$\begin{aligned} [E]_0 &= \frac{[ES]}{K_1[S]} + [ES] + K_2[E][I] \\ &= \frac{[ES]}{K_1[S]} + [ES] + \frac{K_2[ES][I]}{K_1[S]}, \end{aligned}$$

whence

$$\begin{aligned} [ES] &= \frac{[E]_0}{1/K_1[S] + 1 + K_2[I]/K_1[S]} \\ &= \frac{K_1[E]_0[S]}{1 + K_1[S] + K_2[I]}. \end{aligned}$$

Since the velocity of the reaction is $V = k_2[ES]$, we find that

$$V = \frac{k_2 K_1 [E]_0[S]}{1 + K_1[S] + K_2[I]}. \tag{1.9}$$

Assuming that there is a great excess of substrate so that it does not change appreciably during the initial stages of the reaction, we can take S to be constant and so write the relationship between V and I in the form

$$V = \frac{a}{b + K_2[I]}, \tag{1.10}$$

where $a = k_2K_1[E]_0[S]$ and $b = 1 + K_1[S]$. This simple expression allows us to see the effect of different concentrations of a competitive inhibitor such as malonic acid on the initial velocity of the oxidation reaction catalysed by succinate dehydrogenase according to the model described by equations (1.6), (1.7) and (1.8). The graph of this relationship is shown in Figure 1.4.

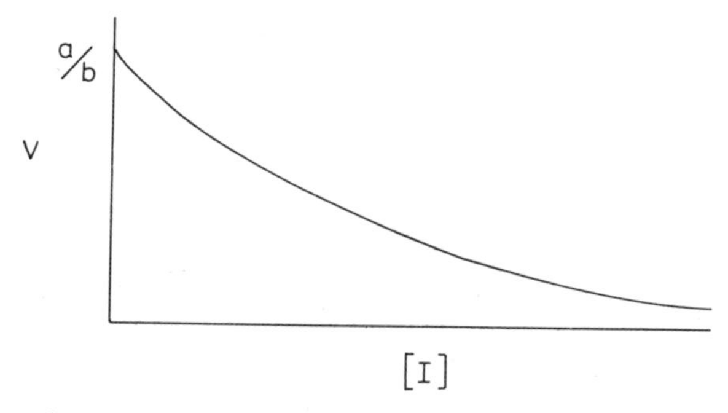

Figure 1.4. The velocity of an enzyme-catalysed reaction, V, as a function of the concentration of a non-competitive inhibitor, I.

This is the inverse image of the familiar Michaelis–Menten relationship between substrate and enzyme reaction velocity, obtained from expression (1.9) by putting $K_2 = 0$, whose graph is shown in Figure 1.5. Both curves

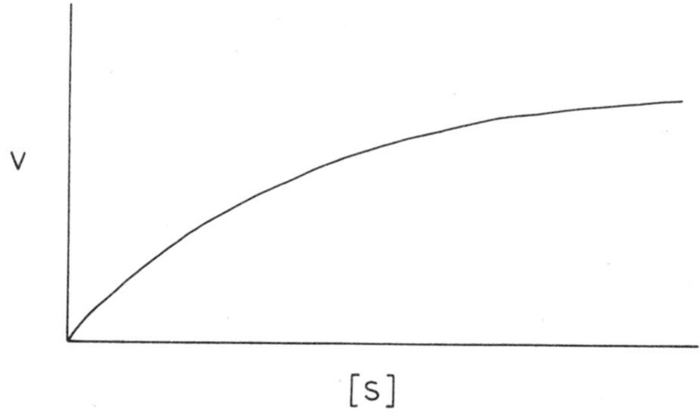

Figure 1.5. The velocity of an enzyme-catalysed reaction, V, as a function of substrate concentration, S, according to Michaelis–Menten kinetics.

show saturation properties as either I or $S \rightarrow \infty$, as is characteristic for surface adsorption or ligand-binding phenomena of the type postulated to underlie macromolecular catalysis. Non-competitive inhibitors such as heavy metals or chelating agents act by a general and non-specific interference with enzyme functions. Their kinetic effects are essentially similar to those described for competitive inhibitors, involving no new principles of enzyme behaviour and giving a relationship of the same type as that shown in Figure 1.4 between enzyme velocity and inhibitor concentration.

End product inhibition, however, involves a new principle of enzyme action because the inhibitor is specific in its reaction with the enzyme (e.g., only CTP and close steric analogues will inhibit aspartate transcarbamylase) but it is sterically unlike the substrate of the enzyme. This is evident from a glance at the substrate and end product inhibitor for aspartate transcarbamylase, shown in Figure 1.6. The new property of enzymes underlying

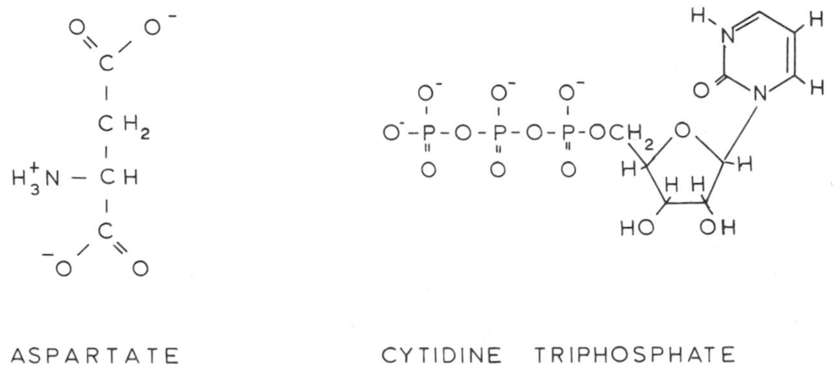

ASPARTATE CYTIDINE TRIPHOSPHATE

Figure 1.6. Aspartate, one of the substrates for aspartate transcarbamylase, is structurally very different from the end product inhibitor, cytidine triphosphate.

their ability to recognize two or more specific steric classes of metabolite was named by Monod and Jacob (1961) allosteric behaviour, for obvious etymological reasons. It was not long before intensive investigation of this property revealed that enzymes are very much more complex entities than had been previously supposed, necessitating a complete revision of Michaelis–Menten kinetics and stimulating the construction of new models of enzyme action.

The experimental studies which underlay these developments started with the isolation and purification of allosteric enzymes from *E. coli* and an investigation of their kinetic behaviour. Aspartate transcarbamylase (ATCase) was studied in this way by Gerhardt and Pardee (1963), while

Changeux (1963) carried out investigations on threonine dehydrase. The first thing to emerge from this work was the observation that the purified, native enzymes no longer showed classical Michaelis–Menten kinetics. When the reaction velocity of ATCase was determined as a function of aspartate concentration, the result was as shown in Figure 1.7, middle curve

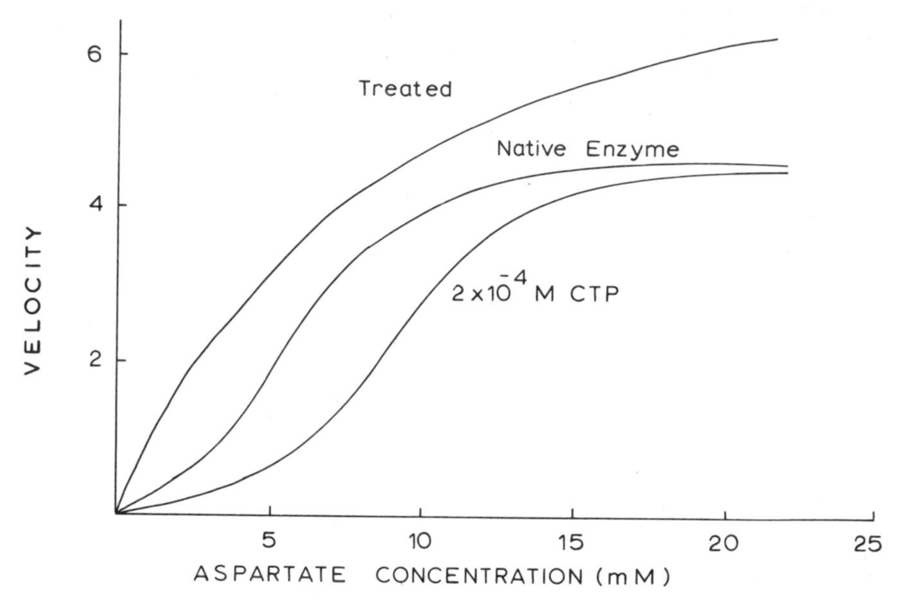

Figure 1.7. The velocity of the reaction catalysed by aspartate transcarbamylase as a function of aspartate concentration under different conditions, as described in the text.

(threonine dehydrase behaves in a qualitatively similar manner). Instead of the curve being a section of a parabola, as it is in Figure 1.5, we have a curve with a point of inflexion. This sigmoid curve, as it is called, already tells us that the kinetics of these enzymes are more complex than can be represented by any model with a single catalytic site, but the other properties discovered for these allosteric enzymes are even more dramatic and require the development of a fundamentally new kinetic theory. It was shown by Gerhardt and Schachman (1965) that if ATCase is treated in a variety of ways such as heating to 62 °C for 10 min followed by rapid cooling on ice, exposure to urea, guanidine or *p*-mercuribenzoate, all of which disrupt hydrogen bonds, the resultant enzyme behaves kinetically according to the Michaelis–Menten scheme, as shown by the upper curve of Figure 1.7. This loss of sigmoid kinetics thus appeared to be due to the dissociation of the enzyme into subunits, a supposition which was confirmed by the isolation of catalytic subunits from the treated enzyme. However, this dissociation had a further consequence, which was a loss of sensitivity of the enzyme to CTP, the end product inhibitor. The native enzyme behaves in the presence of 10^{-4} M CTP as shown by the lower curve of Figure 1.7. It has the same kinetics as the native enzyme, but the substrate concentration required to give one-half maximal velocity, referred to as $[S]_{0 \cdot 5}$, is increased to 10 mM from 6 mM, giving a measure of the amount of inhibition. After disaggregation of the enzyme, however, the catalytic subunits no longer respond to CTP.

Fractionation of the subunits together with crystallographic studies revealed a very remarkable degree of complexity of the enzyme. The regulatory subunit was found to consist of two polypeptide chains (Rosenbusch and Weber, 1971), while there are three chains in the catalytic subunit (Nelbach *et al.*, 1972). A consistent structural model of the native enzyme was then proposed by Cohlberg, Piget and Schachman (1972) in which the catalytic polypeptides are organized into two sets of trimers with equilateral triangular symmetry, the regulatory polypeptides being arranged in pairs around the catalytic units to give a second, larger equilateral triangle. This whole structure depends for its stability on the presence of certain transition elements, zinc being of primary importance. So we see that the native enzyme is a complex entity, a highly organized macromolecular assembly with specific recognition sites on separate monomers for different molecules. The behaviour of the whole is clearly different from the behaviour of its separate parts, so that one is led to postulate an interaction between these parts in seeking an explanation of native enzyme kinetics. What is involved in these phenomena is molecular pattern recognition whereby sites or specific group distributions on the macromolecule interact with particular metabolites or ligands; and the properties of these sites are altered by the binding of other ligands to other sites on the same macromolecule. Interaction between different parts of the same macromolecule are usually interpreted as conformational changes, this being a convenient geometrical representation of what is, generally speaking, a redistribution of energy in the macromolecule.

Co-operative Enzyme Kinetics

Let us consider first the problem of explaining the sigmoid kinetics of ATCase. Since we know that the native enzyme has six aspartate binding sites, it seems reasonable to proceed by modifying the classical enzyme scheme simply by increasing the stoichiometry of the reaction between the enzyme and its substrate. Treating the general case of stoichiometry n, we can write

$$E+nS \underset{k_{-1}}{\overset{k_1}{\rightleftharpoons}} ES^n \xrightarrow{k_2} E+nP \tag{1.11}$$

This leads to the differential equation

$$\frac{\mathrm{d}[ES^n]}{\mathrm{d}t}=k_1[E][S]^n-(k_{-1}+k_2)[ES^n].$$

At the steady state we find

$$[E]=\left(\frac{k_{-1}+k_2}{k_1}\right)\frac{[ES^n]}{[S]^n}=\frac{[ES^n]}{K[S]^n}, \quad \text{where } K=\frac{k_1}{k_{-1}+k_2}$$

Using the conservation condition for total enzyme,

$$[E]_0=[E]+[E]+[ES^n], \quad \text{we get}$$

$$[E]_0=\frac{[ES^n]}{K[S]^n}+[ES^n]$$

whence

$$[ES^n]=\frac{[E]_0K[S]^n}{1+K[S]^n}.$$

From equation (1.11) we then get the result that

$$V=\frac{\mathrm{d}P}{\mathrm{d}t}=nk_2[ES^n]=\frac{nk_2[E]_0K[S]^n}{1+K[S]^n}. \tag{1.12}$$

Evidently if $n=1$ we get the Michaelis–Menten expression given by equation (1.9) with $K_2=0$. But what is the form of the curve if $n>1$? The simplest way to find the qualitative features of V as a function of S is to find maxima and minima and any points of inflexion. So we determine the first and second derivatives. Let us write $nk_2[E]_0K=a$, constant, and drop the square bracket notation for concentrations. Then

$$\frac{\mathrm{d}V}{\mathrm{d}S}=\frac{n^2aS^{n-1}(1+KS^n)-n^2aS^nKS^{n-1}}{(1+KS^n)^2}$$

$$=\frac{n^2aS^{n-1}}{(1+KS^n)^2}. \tag{1.13}$$

The values of S which make this 0, corresponding to a minimum and a maximum respectively, are $S=0$ and $S=\infty$, while for values of S in between these the first derivative is positive. This differs from the Michaelis–Menten expression at $S=0$, where the slope is maximal as seen from Figure 1.5. The existence of a single maximum and a minimum already tells us that there must be a point inflexion in between. To find it, we differentiate (1.13), getting

$$\frac{d^2V}{dS^2}=\frac{n^2(n-1)aS^{n-2}(1+KS^n)^2-2n^3aS^{n-1}K(1+KS^n)S^{n-1}}{(1+KS^n)^4}$$

$$=\frac{n^2(n-1)aS^{n-2}(1+KS^n)-2n^3aKS^{2n-2}}{(1+KS^n)^3}$$

$$=\frac{n^2aS^{n-2}[(n-1)(1+KS^n)-2nKS^n]}{(1+KS^n)^3}.$$

This is 0 at the roots of the equation

$$(n-1)(1+KS^n)-2nKS^n=0$$

or

$$(n-1)-(n+1)KS^n=0$$

whence

$$KS^n=\frac{n-1}{n+1}. \tag{1.14}$$

There is clearly only one positive root, hence one point of inflexion for positive S. Now from (1.12) it is evident that half maximal velocity is reached at $KS^n=1$, while (1.14) tells us that the point of inflection is always reached at a smaller value, only approaching $(S)_{0\cdot5}$ in the limit as $n\to\infty$. For $n=6$, which is the value for ATCase, the point of inflexion occurs at $KS^n=\frac{5}{7}$. We can now draw the curve qualitatively as shown in Figure 1.8.

Another important property to establish is the effect of different stoichiometric values, n, on the maximal slope of the curve, which occurs at the point of inflexion. To find this, we substitute (1.14) into (1.13), getting the result

$$\left.\frac{dV}{dS}\right]_{KS^n=(n-1)/(n+1)}=\frac{n^2a[(n-1)/(K(n+1))]^{(n-1)/n}}{[1+(n-1)/(n+1)]^2}.$$

Evidently as $n\to\infty$, so does this expression, which means that the slope approaches the vertical for very large n. This behaviour is shown by the dotted curve in Figure 1.8. An enzyme showing this type of behaviour would act virtually as a metabolic switch, having practically zero activity until the substrate concentration is near $[S]_{0.5}$ and then rising very rapidly to full activity over a very small range of substrate concentration. It may seem unlikely that enzymes could show this type of behaviour, but stoichiometry 8 has now been shown for some enzymes, giving $KS^n = \frac{7}{9}$ which is beginning to approach switch-like behaviour.

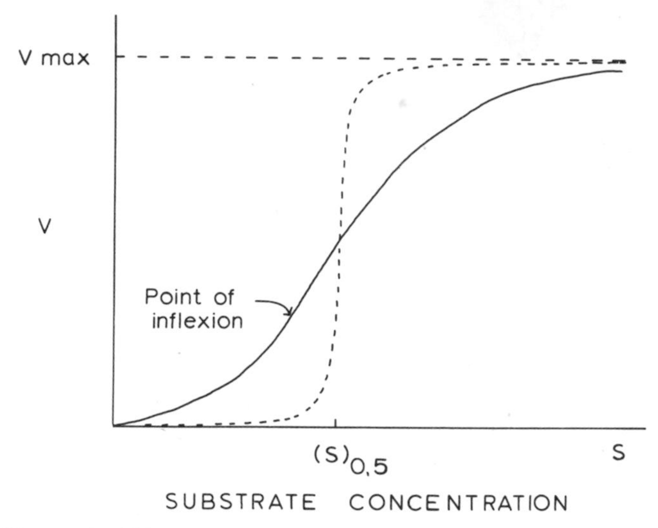

Figure 1.8. The graph of V as a function of S in equation (1.12), with $n > 1$. The solid curve refers to small values of n, and the dotted curve to large values.

At this point we must return to the assumptions of our model and see if we can discover first why a simple change of stoichiometry from the Michaelis–Menten model has such significant consequences; and second, whether this is a plausible model for multimeric enzymes (enzymes with subunits). Another term used to describe sigmoid kinetics is co-operative behaviour. This term is used because up to the point of inflexion, successive small increments of substrate concentration have progressively increasing effects on reaction velocity: i.e. the relation between the variables is non-linear, with positive second derivative. It is as if these successive increments generate a co-operative interaction in the enzyme to facilitate further complex formation and so increase reaction velocity. But we have simply taken $n > 1$, without any assumption about interactions between subunits. Or so it would appear. However, in reality we have made a very strong assumption in writing the reaction in the form of (1.11). We are saying here that complex formation, ES^n, can occur only if n molecules collide simultaneously at the n catalytic sites of the molecule. Such an n-molecular reaction becomes progressively more probable in a non-linear manner as the concentration is increased, until the effects of progressive saturation begin to dominate and the second derivative goes negative. This situation has been referred to as a case of infinite co-operativity in so far as interactions of subunits is concerned, since they must all react with substrate simultaneously or not at all.

Before deciding about plausibility in a purely *a priori* manner, let us see what results have been obtained in fitting this model to experimental data. Although the observation of sigmoid kinetics for enzymes is relatively recent, the essential features of the model described by equation (1.11) were used by A. J. Hill as long ago as 1913 in his classical studies of the oxygenation of haemoglobin. If we write Hb for haemoglobin in place of E in equation (1.11), interpret S to be molecular oxygen and take $k_2=0$, because there is no chemical reaction in the reversible pickup and release of oxygen by haemoglobin, then we get the result

$$[\mathrm{Hb}S^n]=\frac{[\mathrm{Hb}]_0K[S]^n}{1+K[S]^n}$$

The relevant function to look at in this system is the fraction of the total haemoglobin which is oxygenated.

$$N_S=\frac{[\mathrm{Hb}S^n]}{[\mathrm{Hb}]_0}=\frac{K[S]^n}{1+K[S]^n}. \tag{1.15}$$

This of course gives us the same sort of curve as that shown in Figure 1.8. Hill found that in fitting the curve with experimental data it was necessary to take $n=2{\cdot}7$. In terms of the reaction scheme in equation 1.11, this does not make any sense, for it requires that 2·7 molecules of oxygen react with a molecule of haemoglobin. Clearly this number must be interpreted as an average, there being some haemoglobin molecules carrying only one molecule of oxygen, some carrying four (maximum, since there are four haems per haemoglobin) but most with two or three. But this observation destroys the model, which was based upon the assumption that n molecules (necessarily a whole number) react simultaneously to produce the complex. Evidently different numbers of sites can be occupied on different molecules, so it is necessary to allow for this in our model. But if the subunits can react independently with ligands, what causes the sigmoid kinetics, which shows that there is an interaction between the subunits? Evidently this situation demands a model which lies somewhere between the Michaelis scheme, which describes the behaviour of independent units (one site reacting at a time, without any interaction with others) and the Hill scheme of infinite co-operativity.

At this point enzyme kinetics begins to become relatively complex. I will describe the simplest of the models that has been advanced to explain co-operative behaviour and allosteric effects, which is that of Monod, Wyman, and Changeux (1965), and of this only the essence. Pictorially the basis of this model is shown in Figure 1.9, which describes the particular case

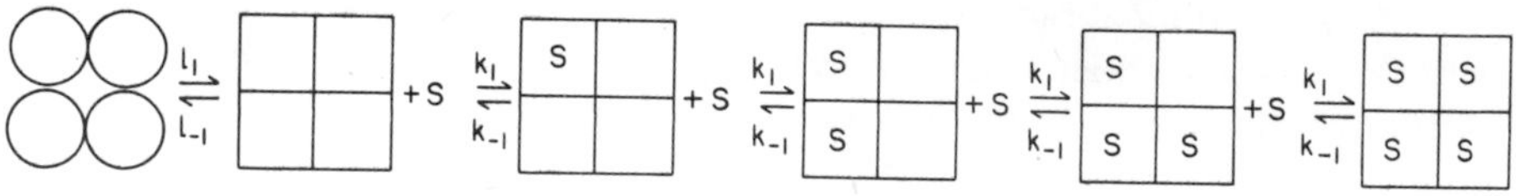

Figure 1.9. A simplified version of the Monod–Wyman–Changeux model of co-operative enzyme kinetics.

of an enzyme with four subunits, each of which has one active site. It is assumed that there are two conformationally distinct states of the enzyme, P and E which undergo spontaneous transitions with rate constants l_1 and l_{-1} as shown. Only state E is capable of reacting with substrate, but then each subunit can react independently, the rate constants being the same for each of these reactions. (Monod *et al.* (1965) allowed both forms of the enzyme, P and E, to react with substrate, but with different affinities. To simplify the argument, one of these has here been taken to be zero.) Now from the point of view of a substrate molecule, each active site is indistinguishable from any other, so all possible combinations of site occupancy will occur without discrimination between them. I have shown only one each of the possible combinations for ES_1, ES_2, and ES_3. In deriving the kinetics of this system, we proceed stepwise as follows. Making a steady state assumption, the relationship between E and P is simply

$$[E]=L[P], \qquad \text{where } L=\frac{l_1}{l_{-1}}$$

The relationship between E and ES_1 must take account of the concentration of active sites available for reaction with S, which is $4E$. Therefore we have the relation

$$4E+S \underset{k_{-1}}{\overset{k_1}{\rightleftharpoons}} ES_1$$

whence

$$[ES_1]=4K[E][S] \tag{1.16}$$

where

$$K=\frac{k_1}{k_{-1}}.$$

For the next step, the concentration of active sites is $3[ES_1]$, one site being occupied; while for the back reaction from ES_2 to ES_1, the concentration of occupied sites is $2[ES_2]$. Therefore we get the relation

$$2[ES_2] = 3K[ES_1][S]$$

whence

$$[ES_2] = \tfrac{3}{2}K[ES_1][S].$$

Substituting for $[ES_1]$ from (1.16) we get

$$[ES_2] = \frac{3 \times 4}{2} K^2[S]^2[E] = 6K^2[S]^2[E]. \tag{1.17}$$

The same considerations give, for the relation between ES_2 and ES_3, the expression

$$3[ES_3] = 2K[ES_2][S]$$

whence

$$[ES_3] = \tfrac{2}{3}K[S][ES_2]$$

and substituting for $[ES_2]$ from equation (1.17) gives

$$[ES_3] = 4K^3[S]^3[E]. \tag{1.18}$$

Finally, the last pair gives us

$$4[ES_4] = K[ES_3][S]$$

whence

$$[ES_4] = \frac{K}{4}[ES_3][S]$$

and, using equation (1.18),

$$[ES_4] = K^4[S]^4[E]. \tag{1.19}$$

Now the most useful function to express the state of this system is the ratio of the concentration of bound substrate to the total concentration of enzyme. This is the same as the mean number of sites occupied per enzyme molecule and will be designated as N_S. The total concentration of bound substrate is evidently

$$[ES_1] + 2[ES_2] + 3[ES_3] + 4[ES_4] = 4K[E][S](1 + 3K[S] + 3K^2[S]^2 + K^3[S]^3)$$

using equations (1.16) (1.19). The expression within brackets is the expanded form of $(1+K[S])^3$, so this reduces to

$$4K[E][S](1+K[S])^3 \tag{1.20}$$

The total concentration of enzyme is

$$\begin{aligned}[P]+[E]+[ES_1]+[ES_2]+[ES_3]+[ES_4] &= \frac{[E]}{L}+[E]+4K[E][S]+6K^2[E][S]^2+4K^3[E][S]^3+K^4[E][S]^4 \\ &= \frac{[E]}{L}+[E](1+K[S])^4.\end{aligned} \tag{1.21}$$

The function we want is equation (1.20)/equation (1.21) giving us

$$\begin{aligned}N_S &= \frac{4K[E][S](1+K[S])^3}{[E]/L+[E](1+K[S])^4} \\ &= \frac{4K[S](1+K[S])^3}{1/L+(1+K[S])^4}.\end{aligned} \tag{1.22}$$

This expression generalizes, for n sites, to

$$N_S = \frac{nK[S](1+K[S])^{n-1}}{1/L+(1+K[S])^n}. \tag{1.23}$$

Comparing this with the Hill function equation (1.15), we see that the main difference is in the occurrence of all powers of the substrate between 0 and n corresponding to the simultaneous occurrence of all possible combinations of enzyme with substrate. There is also the presence of the term $1/L$. If $1/L=0$ (i.e., $l_{-1}=0$, or no P in Figure 1.9), then the equation (1.22) reduces to $4K[S]/(1+K[S])$, which is simply a Michaelis expression, with saturation value $4(S\to\infty)$. So the presence of the term $1/L$ is essential for sigmoid kinetics. That N_S has a point of inflexion may be shown by the same procedure as that used to study the kinetics of the scheme described by (1.12). But since each active site reacts independently with substrate, wherein lies the co-operativity of this scheme? Our attention has already been drawn to $1/L$ in (1.22) as the source of the sigmoid nature of the curve, so we might suspect that it is the conformational relationship between P and E that is producing the co-operative effect. A basic assumption of this model is that all subunits change their conformational state simultaneously.

Evidently once a substrate molecule has reacted with any site to generate ES_1, this complex cannot revert to P except via E. So the first reaction of substrate with E produces a molecule with three sites for reaction which will be available so long as S is bound to its site. The next step has a similar effect, but less so, since with ES_2 only two sites become available. After this, saturation begins to dominate and the curve approaches its asymptotic upper limit. Thus we see that the origin of co-operativity in this model is the existence of the alternative conformational configurations P and E, together with the stabilizing effect of S on the catalytically active form of the molecule.

So much for sigmoid kinetics in this model. What about allosteric effects? These are assumed to arise via changes in the equilibrium constant L. If this is decreased (e.g. l_1 is decreased) then the function equation (1.22) will rise more slowly with increasing S. This reproduces the effect of feedback inhibition as shown in Figure 1.7. The mechanism of action of the inhibitor could be, then, to stabilize P by binding to allosteric sites which are available in that conformation. Such sites would have to be included in the model and the same argument as that used for substrate would then be used for inhibitors, as is done in the paper by Monod *et al.* (1965) to which the interested reader may refer. One can also imagine ligands which stabilize E, thus increasing L. These would then be activators of the enzyme, ATP having this effect on ATCase. Another well studied example is isocitrate dehydrogenase which is activated by AMP (Atkinson *et al.*, 1965).

A yet more general model of allosteric behaviour in proteins would allow subunits to change conformational state individually, but with some interaction such that the probability of transition depends upon the states of other subunits in the multimer. Such a generalization involves a whole class of possible models, some of which have been studied by Koshland (1970). Their algebra is more complex than the relatively simple expression equation (1.22) and at the present state of the art it is difficult to distinguish experimentally between the different possibilities. Despite this apparent complexity, it is evident from the simple geometry of Figure 1.9 that the conceptual foundations for our understanding of macromolecular behaviour are relatively primitive. The theory presented cannot be regarded as in any sense a physical theory, for there is no expression for the energies of the different conformational states. This simply reflects the difficulty of devising any satisfactory theory of macromolecular order, a situation that is hardly surprising in view of the diversity and the nicety of behaviour of which proteins are capable. They constitute molecular pattern recognizing systems of quite exquisite variety and refinement, which properties confer upon them the ability to play the fine discriminating roles necessary for the regulation of metabolic flow.

Metabolic Control Circuits and their Dynamics

I would like at this point to introduce a symbolic convention for the representation of metabolic control processes which will be useful in building up the hierarchy of regulatory levels that characterize integrated organic behaviour. The picture of Figure 1.1 has been found to be inadequate to accommodate specific regulatory interactions between enzymes and metabolites. I shall now use boxes to represent metabolites and circles to represent enzymes. Thus a metabolic sequence with feedback from the end product to the first reaction in the pathway will be represented as in Figure 1.10. The minus sign attached to the feedback arrow designates a negative

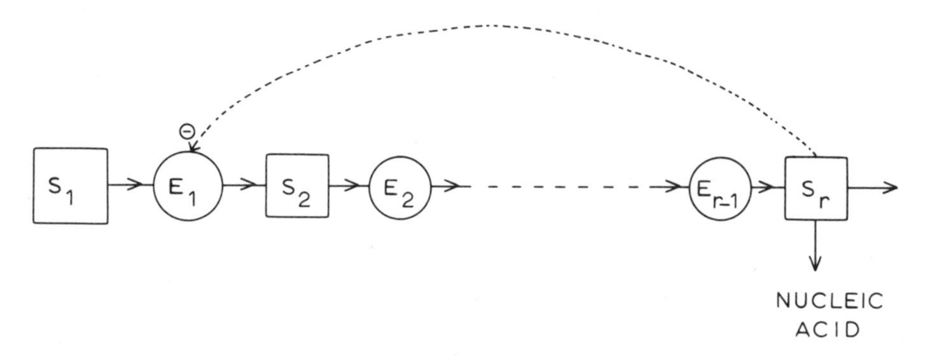

Figure 1.10. A schematic description of a metabolic pathway with feedback inhibition of the first enzyme in the pathway, E_1, by the end product, S_r, which is a precursor for nucleic acid synthesis.

effect of the end product on the enzyme. Comparing this representation of a metabolic pathway with that in Figure 1.1, it is evident that the specific feedback loop creates a partially autonomous element of control within the total metabolic system, which thus becomes partitioned into physiological regulatory units each of which has a degree of independence in its behaviour. Evidently under different conditions of nutrition, one or other of these units could become rate limiting or controlling in relation to the whole, so that there is no dominant unit. The functional relationships of these units are flexible and variable, but stable overall, serving the demands of physiological adaptation at this level of cellular organization. Such a constellation of interacting, partially autonomous but integrated units with well-defined behavioural routines (sub-routines of the whole) has been referred to as a heterarchy, in distinction to the pyramid-like structure of a hierarchy which has rigidly-defined dominance relations between its constituents. We will find that both heterarchy and hierarchy play important roles in physiological organization and that temporal considerations determine which aspect of behaviour is most in evidence.

Having established that a metabolic network of the type depicted in Figure 1.1 is in general very stable, its variables returning to steady state values within a few minutes of a disturbance, it is important to enquire into the stability characteristics of a system consisting of many interacting elements of the type shown in Figure 1.10. In general it is widely recognized that negative feedback, while conferring overall stability on a system, may nevertheless result in continuous oscillations of the variables about some mean value. Thus the discovery of feedback inhibition changes in a significant manner the dynamic behaviour that one might anticipate in the metabolic system. And in the past 10 years or so this changed expectation has resulted in some very interesting experimental and theoretical observations about oscillatory behaviour in cellular metabolism.

There are two general dynamical consequences of negative feedback in kinetic systems. One is an increase in stability over that provided by mass action, which of course still operates in any metabolic system; and the other, paradoxically, is a decrease in stability, because of the possibility of autonomous oscillations. However, these oscillations are themselves stable, being referred to as limit cycles. So really what arises with specific feedback control circuits is a higher-order property: dynamical stability in relation to limit cycles, so that stable rhythmic behaviour emerges to supplement the more familiar stability in relation to a point.

If a system with negative feedback is point-stable (stable in relation to a steady state), then the feedback terms will tend to 'stiffen' the resistance of the system to a perturbation, for an active response is present. One can see this in descriptive terms in Figure 1.10. Suppose the disturbance is such as to decrease S_r. Then by mass action, there will be a reduced outflow of S_r from the pool; but in addition to this, there will be an increased flux rate down the pathway as a result of the reduced inhibition of E_1, and so an increased rate of production of S_2 from precursor. This direct effect of a metabolite on a rate is what changes the dynamic picture in an important respect when feedback effects occur. The system then responds actively to disturbances, resisting them if feedback is negative; but if this active response is delayed in time, then oscillations can result. Thus we see that the situation becomes dynamically rather more complex and at the same time, biologically speaking, considerably more interesting.

To explore the theoretical side of this behaviour first, let us consider a simple two-step control circuit where $r = 3$ in Figure 1.10. Let us take S_1 to be a constant source of substrate for E_1 and suppose that E_2 reacts with S_2 according to the scheme (1.11). From the point of view of stability and oscillation, only the highest power of the reactants will be significant, so that little would be altered by using the kinetically more plausible scheme leading to equation (1.23) in place of the Hill formulation. Thus we use a Hill-type formulation for the inhibitory interaction of S_3 with E_1,

$$E_1 + m\,S_3 \underset{k_{-3}}{\overset{k_3}{\rightleftharpoons}} E_1S_3^m \tag{1.24}$$

where m is the stoichiometry of the reaction leading to the inhibited, inactive complex $E_1S_3^m$. The differential equations for S_2 and S_3 then take the form

$$\left.\begin{aligned} \frac{\mathrm{d}S_2}{\mathrm{d}t} &= \frac{a}{b + K_3[S_3]^m} - \frac{c[S_2]^n}{1 + K_2[S_2]^n} \\ \frac{\mathrm{d}S_3}{\mathrm{d}t} &= \frac{c[S_2]^n}{1 + K_2[S_2]^n} - d[S_3] \end{aligned}\right\} \tag{1.25}$$

where

$$a = pk_1{}'K_1[E_1]_0[S_1]^p,$$

and

$$b = 1 + K_1[S_1]^p,$$

both constant on the assumption that S_1, interacting with enzyme E_1 with stoichiometry p, is a constant; $c = nK_2[E]_2$, and d is the rate of removal of the end product S_3, from the pool. In general, equation (1.25) will have a single positive steady-state solution obtained by putting $\mathrm{d}S_2/\mathrm{d}t = \mathrm{d}S_3/\mathrm{d}t = 0$, although multiple steady states can exist. The question of stability can be investigated by linearizing the equation in the neighbourhood of a steady state, designated by $\bar{S}_2$, $\bar{S}_3$. This is done by defining new variables, $U_2 = S_2 - \bar{S}_2$, $U_3 = S_3 - \bar{S}_3$, and then expanding the non-linear functions to the first power in the variables U_1, taken to be small. In terms of the new variables, equation (1.25) takes the form

$$\begin{aligned} \frac{\mathrm{d}U_2}{\mathrm{d}t} &= \frac{a}{b + K_3(\bar{S}_3 + U_3)^m} - \frac{c(\bar{S}_2 + U_2)^n}{1 + K_2(\bar{S}_2 + U_2)^n} = F_1(U_3) - F_2(U_2) \\ \frac{\mathrm{d}U_3}{\mathrm{d}t} &= \frac{c(\bar{S}_2 + U_2)^n}{1 + K_2(\bar{S}_2 + U_2)^n} - d(\bar{S}_3 + U_3) = F_2(U_2) - d(\bar{S}_3 + U_3). \end{aligned} \tag{1.26}$$

The first derivatives of the functions are

$$\left.\frac{\mathrm{d}F_1}{\mathrm{d}U_3}\right]_{U_3=0} = \frac{-maK_3[\bar{S}_3)^{m-1}}{[b + K_3(\bar{S}_3)^m]^2} = -\alpha$$

and

$$\left.\frac{\mathrm{d}F_2}{\mathrm{d}U_2}\right]_{U_2=0} = \frac{nc[\bar{S}_2)^{n-1}}{[1 + K_2(\bar{S}_2)^n]^2} = \beta.$$

Then equations (1.26) reduce to

$$\frac{dU_2}{dt} = F_1(O) - \alpha U_3 - F_2(O) - \beta U_2$$

$$= -\beta U_2 - \alpha U_3$$

$$\frac{dU_3}{dt} = F_2(O) + \beta U_2 - d(\bar{S}_2 + U_3)$$

$$= \beta U_2 - dU_3,$$

since $F_1(O) = F_2(O) = d\bar{S}_3$ by the steady-state assumption.

The stability of these linear equations depends upon the roots of the characteristic equation,

$$\begin{vmatrix} -\beta - \lambda & -\alpha \\ \beta & -d - \lambda \end{vmatrix} = (\beta + \lambda)(d + \lambda) + \alpha\beta$$

$$= \lambda^2 + (\beta + d)\lambda + \beta(\alpha + d) = 0.$$

Since all the constants are positive quantities, this equation cannot have any roots with non-negative real parts. Thus the system cannot oscillate and it is stable in the neighbourhood of the steady state. Evidently, then, negative feedback cannot alter the stability of a two-step reaction sequence. However, one more step in the sequence before feedback occurs is enough to allow instability and the occurrence of a limit cycle oscillation providing $m > 8$, as shown by Walter (1969). His detailed analysis of what he calls Yates–Pardee feedback loops show the exact conditions under which such control circuits can show oscillatory behaviour, which depends critically on both the number of steps (value of r) and the stoichiometry of the controlled reaction (value of m). Walter used linear expressions for enzyme kinetics, thus ensuring the existence of a single positive steady state solution. An analytical treatment of the relation between the number of steps in the reaction sequence and the stoichiometry of the feedback inhibition required to give instability has been presented by Viniegra–Gonzalez and Martinez (1969) and in more rigorous form in a recent interesting and valuable study by Rapp (1975).

As expected from the relaxation time arguments of pages 2 and 3, the periods of oscillation in feedback loops when they occur are of the order of a few minutes, as calculated on theoretical grounds (Goodwin, 1963) or from computer simulations when appropriate parameter values are used. Only relatively recently have soluble *in vitro* enzyme systems been observed to show stable oscillatory behaviour of the type predicted. Yamazaki *et al.* (1967) obtained oscillations in an open system using horse-radish peroxidase and lactoperoxidase in the presence of a reductant and oxygen, the period being about five minutes. This system has a feedback structure whose analysis has been presented by Degn and Mayer (1969). A number of such systems have now been studied, but perhaps the most interesting is the glycolytic oscillator, first reported in whole cells by Duysens and Amesz in 1957 and later reconstructed *in vitro* and studied intensively in the laboratories of B. Chance in Philadelphia and B. Hess in Dortmund (see review by Hess and Boiteux, 1971). Cell-free extracts of yeast undergo oscillations in all glycolytic intermediates with periods in the range of 2–20 min in response to the addition of an appropriate carbon source. This is a complex system, but detailed analysis has revealed that one of the enzymes primarily responsible for the oscillatory behaviour is phosphofructokinase (PFK). This enzyme is a major control point in glycolysis, being activated by a product, adenosine monophosphate, and inhibited by the end product of glycolysis, ATP. In terms of the conventions of Figure 1.10, the relevant part of the glycolytic sequence for oscillatory behaviour can be shown as in Figure 1.11. Here S_{11} = fructose 6-P; S_{12} = ATP as substrate; E_1 = PFK;

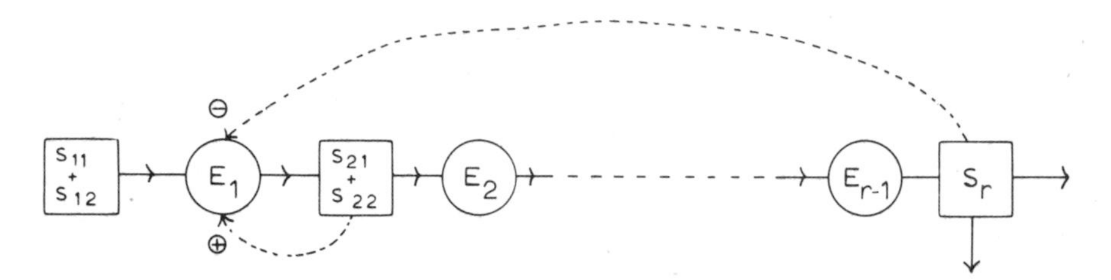

Figure 1.11. A metabolic pathway with positive and negative feedback loops, as occurs in the glycolytic sequence.

S_{21} = fructose 1,6-diP; S_{22} = ADP (in equilibrium with AMP); E_2 = aldolase; S_r = ATP. We know that the negative feedback loop is sufficient to generate oscillations under certain conditions; but it is clear that the existence of the positive feedback loop will increase the instability of the system, making oscillations more probable. Indeed, it is not difficult to show that if $r = 3$ so that negative feedback occurs after two steps, then oscillations can occur if the positive feedback is present. A full simulation of the glycolytic system has been presented by Higgins (1967). The universality of glycolysis in cells suggest that there may be some very significant physiological role for this oscillating system, but this has as yet proved elusive. I will consider a possible embryological role for this kind of metabolic oscillator in a later chapter.

Another physiologically very intriguing case of metabolic oscilliations occurs in mitochondria, where the whole organelle undergoes periodic spatial and biochemical changes under well defined conditions. Volume, ion transport, respiration rate and redox state of cytochrome *b* are among the variables which cycle with periods of about one minute. No explanation has been provided for the function this might play.

CONCLUSION

In moving from the representation of metabolic networks in the form of Figure 1.1 to an interacting set of elements such as those shown in Figures 1.10 and 1.11, two noticeable features have emerged, one a simplification and the other a complication. The simplfying feature is the identification of partially autonomous sub-systems of the whole network, consisting of the metabolic sequences which lead to particular end products. The activity of such sequences, measured as the rate of transformation of the component metabolites, can be altered independently of other parts of the total system by the level of the end product acting by negative feedback on the first enzyme. Such a partitioning of the system into components which can respond independently to perturbations confers great adaptive flexibility on the whole. It also increases the response rate over that which would occur in a totally connected network of the type first considered. At the same time one has the added dynamical complication that the feedback circuits can result in continuous oscillations in the concentrations of metabolites with periods in the minute range. This, of course, provides a richer range of dynamic behaviour for the cell to exploit in seeking maximally adaptive strategies; it is only from the analytical viewpoint that difficulties arise, since we have few mathematical tools to cope adequately with non-linear systems of the type described by equation (1.25), leaving aside the additional complications of possible positive feedback loops and other interactions. While it is clear that computer solutions of such equations can lead to important qualitative and quantitative insights into the behaviour of the systems, as in the work of Walter referred to, one continues to hope for the emergence of procedures which might allow a somewhat more perspicuous treatment of the whole of non-linear dynamics. In the last chapter I shall discuss one such approach which seems to me to be promising.

Chapter 2

STABILITY AND REGULATION IN THE EPIGENETIC SYSTEM

Enzymic Adaptation

SINCE Pasteur's historic discoveries about micro-organisms and his work on the fermentation of wine, the study of microbial physiology has flourished in France. The first systematic and comprehensive account of the phenomenon of enzymic adaptation is to be found in a book by Duclaux published in 1899, entitled *Traité de Microbiologie.* It is therefore particularly appropriate that research on this aspect of cellular physiology should have led in the late fifties and early sixties to one of the most successful theories of molecular biology, fruit of the work of F. Jacob, J. Monod, and their colleagues at the Pasteur Institute in Paris.

We saw in Chapter 1 that the first evidence of the existence of specific control processes in bacterial metabolism emerged as a result of studies with mutant strains, combining the analytical power of genetics and biochemistry. The elucidation of the mechanism of bacterial adaptation was likewise dependent upon these techniques applied with great subtlety and imagination to this problem by the investigators at the Pasteur Institute. It is interesting to note that a detailed review of what was then known about enzymic adaptation was published by Monod in 1947, a year after the paper by Lederberg and Tatum (1946) which announced the discovery of the occurrence of sexual recombination in *Escherichia coli.* In this review Monod deduced from the evidence then available regarding mutant strains that the 'adaptive potentialities' of cells may undergo either very important or slight mutative variations affecting a single enzyme, without in the least the 'potential properties' of other enzymes attacking closely related, mutually interacting substances being affected. He suggested that 'the specific properties of each enzyme are determined by an *independent* hereditary factor'. He then continues: 'It should be realized, however, that this evidence remains circumstantial and incomplete, as long as straight genetical tests have not been applied. It is hoped that Lederberg and Tatum's brilliant discovery might make such tests possible'. A little more than a decade later, Jacob and Monod had succeeded in unravelling the molecular details underlying gene control processes in bacteria, capitalizing on the use of mutants and recombinants to prove decisively the genetic properties anticipated by Monod and to go well beyond into the new world of transcription, translation and the molecular principles of prokaryotic regulation.

The basic facts of adaptation are very easily stated. If a culture of a bacterium such as *Escherichia coli* growing on a minimal salt medium is provided with a mixed carbon source such as glucose and lactose, for example, then initially they grow exponentially and then, quite suddenly, they plateau as shown in Figure 2.1. This break in the growth curve lasts for

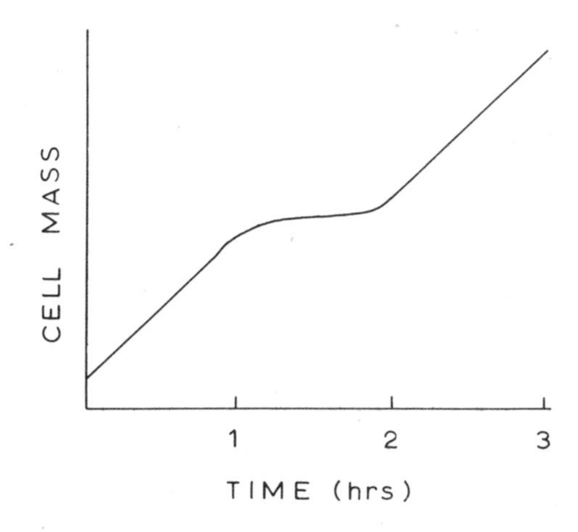

Figure 2.1. Diphasic growth of a bacterial culture on a mixed carbon source such as glucose and lactose. During the initial exponential growth phase, only glucose is utilized, while during the second exponential phase, only lactose. Adaptation occurs during the intervening phase of no growth.

some 50 to 60 minutes and then growth resumes at a rate virtually identical with the initial growth rate.

It is easily shown that during the second growth phase an enzyme is present which was absent during the initial phase. This is β-galactosidase, whose role is to cleave the β-galactosidic bond of lactose to give the hexoses, glucose and galactose. In fact there are two other enzymes which also appear during the adaptation process, β-galactoside permease and β-galactoside acetylase, both of which are concerned with the transport of lactose across the cell membrane into the cell interior where it can be catabolized. Evidently during the initial growth phase, the cells grow exclusively on glucose, for lactose cannot enter the cell until the enzymes required for its transport are synthesized. Only after the glucose supply has been exhausted, as can easily be shown experimentally, do the cells begin to adapt to lactose utilization by synthesizing the three enzymes. These observations raise two basic questions relating to enzyme adaptation: (i) what are the molecular processes involved in the induction of new enzymes in a cell under defined changes in the environment; and (ii) what prevents these enzymes from being synthesized when glucose is present even though the inducer (lactose) is available?

It is immediately obvious from the adaptation time that we are dealing with a process which is slower than a change of state in the metabolic system, as defined in the last chapter. This observation alone makes it unlikely that enzymic adaptation involves a simple activation of a previously existing enzyme present in the cells in an inactive form, since we saw in the last chapter that such interactions must occur in seconds or less. However, it was necessary to confirm this independently by methods such as those described by Spiegelman (1948), who showed that supernatant fractions from boiled extracts of adapted yeast were unable to activate protein extracts from unadapted yeast; and by immunological methods as described by Cohn and Torriani (1952), who showed that there is no protein in unadapted bacteria which cross-reacts with antibody prepared from purified β-galactosidase. These and other investigations such as incorporation of radioactive amino acids into β-galactosidase during induction established that the enzymes were synthesized *de novo* during adaptation. But in order to be certain that individual cells were changing their physiological state from the unadapted to the adapted condition, it was necessary to exclude the possibility that genetic selection was occurring in the population in response to the change of nutrient conditions. This had in fact been established by Monod (1947), but again a simple consideration of time scale is sufficient to establish a strong *prima facie* case against the selection of mutants. Bacteria growing on a medium of minimal salts plus a carbon source such as glucose or lactose have a generation time of around 50 minutes, about the same time as that required for enzyme induction. Thus only one generation of a mutant strain could be selected in this time and a resumption of culture growth at the same rate as that occurring prior to the induction could not possibly occur if only the selected mutant strain was contributing to growth. We are using here the same kind of argument as that used in the previous chapter to identify the metabolic system by considerations of rates of change, bounding metabolic rates below by enzyme turnover numbers and above by times required for changes in enzyme concentrations. Now we are identifying the next system in the hierarchy, that involved in enzyme adaptation, bounding it below by the metabolic system which includes phenomena such as enzyme activation and inhibition; and bounding it above by genetic selection, which must take several generations. Enzymic adaptation occurs within a single generation time and so is not a genetic phenomenon. It is epigenetic in the sense that it involves changes in gene activities rather than gene mutation and selection.

So we may refer to the system we are now focussing our attention upon as the epigenetic system of the cell. Evidently its response time is of the order of several minutes to one hour. The response time of the genetic system must be several generations, which in micro-organisms will be several hours. For organisms with longer generation times, the response or characteristic times of the different systems will be altered correspondingly. And we will find that for more complex organisms the epigenetic system involves many processes besides gene induction and repression, but these are always included as an essential feature of epigenesis.

Gene Induction and Repression

The observation that the three enzymes involved in the metabolism of lactose (β-galactosidase, β-galactoside acetylase and the permease), were always collectively present or absent and occurred in fixed proportions in wild type cells suggested that whatever the nature of the control process, it acted in a unitary or co-ordinated manner on the three genes involved in specifying the three enzymes. The collective term for the whole control system is the *lac* operon, a unit of epigenetic control whose etymology arises from the idea of a system controlled by an operator. As stated above, the key to an unlocking of the mechanism of adaptation was provided by the use of genetic methods. Mutant strains of *E. coli* were found which produced β-galactosidase and the other two enzymes continuously, irrespective of the presence of glucose or the absence of lactose. These were called constitutive mutants, cells which had lost the property of inducibility by lactose. They could then be conjugated with wild type cells to give partial diploids heterozygous for inducibility, whose behaviour could then be studied. It was found that the original mutants which had the phenotypic characteristic of constitutivity (always made the enzymes, irrespective of the nutrient conditions) fell into two classes, in one of which the heterozygote behaved like the wild type cell, thus being inducible, while the other continued to behave constitutively. The first class of mutant was labelled i^-, indicating that the property of inducibility by lactose was defective in some way. The heterozygote $(i^{+\prime}i^-)$ then has wild type characteristics, the wild type gene (i^+) being dominant.

The other type of constitutive was labelled O^-, indicating that there was some deficiency in the operative part of the control system. This defect is not corrected by the insertion of a wild type gene (O^+) into a cell carrying O^-. Genetic mapping studies showed that these two sites were distinct, but not far apart on the *E. coli* chromosome. Furthermore, it was found that the O^+ gene maps very close to the structural genes for the three enzymes, while the i^+ gene is a little further away (higher frequency of recombination).

The dominance of i^+ over i^- immediately suggests that the i^+ gene produces some molecule which can diffuse through the cell and correct the deficiency of operation of the other chromosome. The O^+ does not exert any action at a distance of this sort. Indeed, it was found that an O^- affects only the structural genes on the same chromosome, allowing them to produce the three enzymes involved in lactose catabolism. Thus O^- expresses itself only in the *cis*-position. If a chromosome carrying wild type *lac* operon genes is present in a cell with an O^- mutation on the other chromosome, then both behave independently. Such a partial diploid will produce *lac* operon enzymes from the mutant operon in the absence of lactose; and it will produce a second lot of *lac* operon enzymes from the normal operon when induced with lactose. These actions are not quite independent because of a complication which we shall shortly consider, but they are nearly so.

On the basis of this evidence, Jacob and Monod (1961) proposed a model which accounted for the observations in a very elegant and economical manner, and made some important predictions. Their picture was essentially as shown in Figure 2.2. Here the line represents part of the bacterial

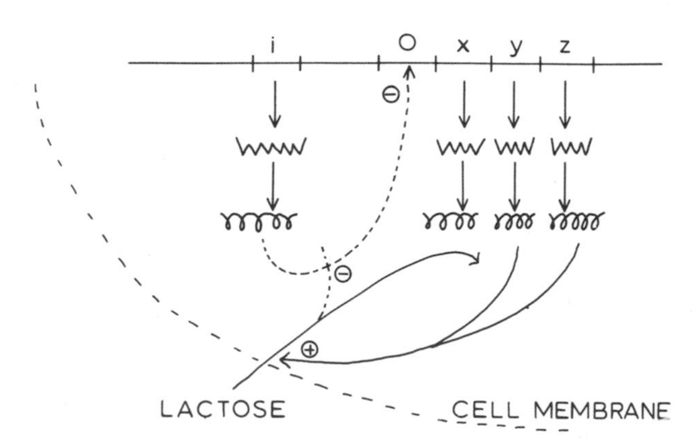

Figure 2.2. A schematic description of the *lac* operon as initially proposed by Jacob and Monod (1961). For details, see the text.

chromosome with loci labelled *i*, *O*, and *x*, *y*, *z* for the three structural genes *β*-galactosidase, acetylase and permease, respectively. The dotted line is the cell membrane, and a negative sign in a circle indicates inhibition or repression. The *i* gene, referred to now as the regulator gene, produces a messenger RNA which is translated into a protein, the repressor. This protein diffuses to the site *O*, the operator gene, and when combined with this locus blocks (represses) the transcription of mRNA from the structural genes *x*, *y* and *z*. This is presumed to occur by a covering of the initiation site for RNA polymerase action. Lactose inactivates the repressor by a ligand-binding reaction of the type considered in the last chapter, altering the three-dimensional configuration of the repressor and rendering it unable to combine with *O*. Thus according to this model, induction is actually *depression*: release of the structural genes from repression. Evidently a defect in the *i* gene, giving a faulty repressor which is unable to combine with *O*, will give a constitutive mutant which will be corrected by the presence of a wild type i^+ in the same cytoplasm. A mutant *O*, on the other hand, will give a constitutive which cannot be corrected by a normal *O* gene, and the *cis*-position effect is explained.

The essential features of this control model have been spectacularly verified since its publication, and many other details have been added to complete the picture. The postulated repressor protein was isolated (Gilbert and Müller-Hill, 1966) and its behaviour is just that predicted by the model. It has been found to be a tetrameric protein with four identical subunits, each capable of binding one molecule of inducer. It is not yet clear whether lactose itself interacts with this repressor in the normal operation of the induction process or whether an early product of lactose catabolism, such as allo-lactose is the inactivating molecule. If the latter, for which there is some evidence (Burstein *et al.*, 1965) then the induction process would involve a positive feedback loop: allo-lactose, a product of β-galactosidase activity, inactivates the repressor and so results in more β-galactosidase production. But there is another positive feed-back loop which brings our attention to an interesting anomaly in this model. The other two enzymes are responsible for the transport of lactose into the cell. The more lactose, the more is the repressor inhibited, and the more transport enzymes will be produced. This is a runaway process which will stabilize only when all the repressor has been inactivated and the structural genes are working at full capacity, but the more important question is: how does it get started? One needs some of the transport enzymes in the cell membrane simply to get lactose into the cell interior to start the induction. How is this pump primed? The answer is that the pump is always primed because there is a leak in the control system. An uninduced population of bacteria has, on average, 0.5 molecules of β-galactosidase per cell. This represents basal enzyme synthesis, due to the 'noisiness' of the control mechanism: the repressor must occasionally dissociate from the operator due to thermal noise effects and allow mRNA synthesis to occur briefly. Evidently imperfection is an important feature of cellular design.

The model of Figure 2.2 gives us essentially a switching circuit based on positive feed-back which would turn on in the presence of lactose and off (with some noise) in its absence. But we saw from the discussion of enzymic adaptation that in the presence of glucose the *lac* operon is switched off, and induction by lactose begins only after glucose has been removed from the medium. How is this 'glucose effect' to be accounted for? At this point it is necessary to embed the simple picture of Figure 2.2 into the complexities of cellular metabolism and to consider the physiological role played by the *lac* operon. If the cell is supplied with glucose, there is no need, speaking teleologically, for the *lac* operon to function. It would only add more glucose to an already adequately provided glycolytic pathway, and the production of the enzymes of the *lac* operon would divert nucleotides and amino acids into unnecessary molecules. This is like the situation described in the last chapter where a cell provided with an end product, say isoleucine, has no need to produce the precursors for isoleucine synthesis and so controls their production from the source (first enzyme in the pathway leading exclusively to that end product). Here we saw the operation of negative feed-back, and we might anticipate that a similar process would be involved in controlling the activity of the *lac* operon when a good carbon source such as glucose is available. The problem is to find the end product of carbohydrate catabolism. There are many catabolic products which could play this role and many have been studied in the search for the elusive 'catabolite repressor' as this has been called (Prevost and Moses, 1967). Such a repressor would be expected to control not only the *lac* operon, but any enzyme system which can contribute molecules to the common catabolic pathway for energy and building blocks. This includes many degradative enzymes such as tryptophanase, some deaminases, histidase, some phosphorylases, etc. What we are looking for is some molecule which is involved in the regulation of a wide spectrum of catabolic enzymes, controlling the basic energy flux within the cell. In recent years the evidence has pointed strongly to cyclic AMP as a molecule involved in this role, acting as the 'glucose distress signal' within the cell (Makman and Sutherland, 1965; de Crombrugghe, 1971; Pastan and Perlman, 1970). However, the effect of this molecule is positive in that it induces catabolic enzymes; while in its absence, these enzymes are not synthesized to any great extent. Thus it is necessary to consider a control system in which a high level of glucose will give low levels of cAMP and conversely; and cAMP must play an obligatory inductive role in the operation of the *lac* operon.

An example of a system with these properties is shown in Figure 2.3. I have changed the symbolism from Figure 2.2 to conform with that introduced in the last chapter to describe molecular control processes, using circles to represent control points (actual or potential) and boxes to represent metabolic pools. Here the *G*s represent genes or operons, *R*s are polysomes, P_i is the repressor, E_{l_1} is β-galactosidase, E_{l_2} represents the transport enzymes, and E_a is adenyl cyclase which makes cAMP from ATP. The *lac* operon is represented as a single unit of epigenetic control. Lactose enters the cell via the transporting enzymes located in the cell membrane and then is shown (S_1) as an inactivator of the repressor. The positive feed-back nature of this part of the control circuit is evident, since the two

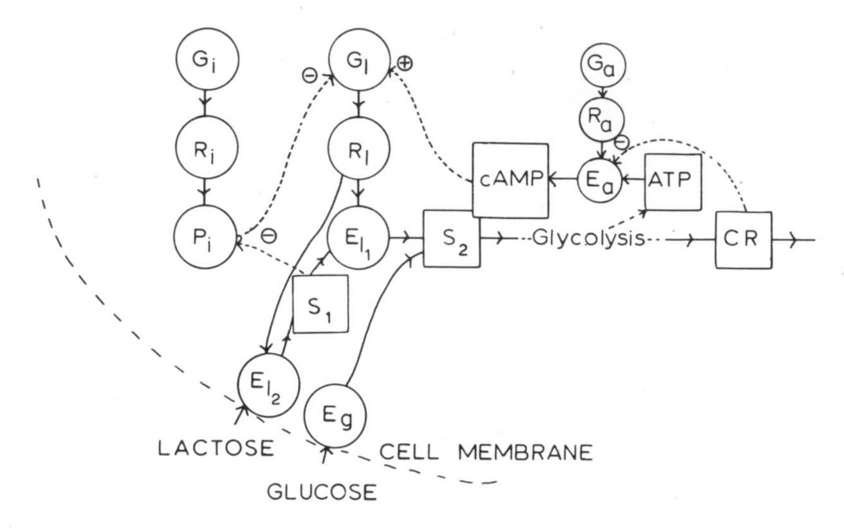

Figure 2.3. A possible control scheme for the *lac* operon, including catabolite repression.

negatives make a positive and the transporting enzymes have a positive effect. The products of β-galactosidase are shown as S_2, which enters the glycolytic pathway and produces the catabolite repressor (CR). Glycolysis also produces ATP, which is converted to cAMP by adenyl cyclase. The catabolite repressor is taken to be an inhibitor of adenyl cyclase, as suggested by the studies of Peterkofsky and Gazdor (1974). Cyclic AMP is assumed to be necessary for message transcription from the *lac* genes. There is evidence that it does play this role at a site called the promoter: part of the *lac* operon which is involved in the attachment of RNA polymerase to the DNA and the initiation of message transcription (Beckwith, 1967). This site is located between the i and the O genes in Figure 2.2.

Also shown in Figure 2.3 is the glucose-transporting enzyme, E_g, with glucose entering the glycolytic pathway at S_2. It is then evident why wild type cells will not produce β-galactosidase (and the transporting enzymes) if glucose is present, since in this case the pool of catabolite repressor would remain filled and little cAMP would be available for transcription initiation. Evidently this role of cAMP must be specific for the genes coding for the catabolic enzymes of the cell, since many other genes are transcribing messages under these conditions. The 'glucose distress' function of cAMP appears to be an example of a control signal functioning at a relatively high level of physiological regulation, since it operates not simply for a single operon, as does lactose, but for a class of operons all subserving the same physiological function, the provision of energy and building blocks for anabolism via catabolic pathways. The general principle of negative feed-back is used to achieve this regulation, but clearly the details of the control interactions can be quite complex. Figure 2.3 is only one of a number of possible circuits.

FEED-BACK REPRESSION OF BIOSYNTHETIC PATHWAYS

Although the *lac* operon was the first gene control system to be analysed in detail, it is not the most typical example of gene regulation in prokaryotic cells. We found in the last chapter that a sufficient condition for stable regulative performance in a control circuit is the occurrence of a negative feed-back loop involving an end product. In the case of the *lac* operon this is provided by catabolite repression. The inductive or derepressive effect of lactose is an additional feature of this system, providing a short, rapidly-acting, positive feed-back loop which results in a quick response of the operon to an inductive stimulus. This response is stabilized by the more slowly-acting repressive effect of the catabolic control signal. This circuitry is reminiscent of the glycolytic control system as drawn in Figure 1.11, and it has the same dynamic properties: it is likely to go into an oscillatory state. I will consider the experimental evidence for such behaviour later. At the moment I wish only to point out that the circuitry of the *lac* operon is somewhat more complex than one which operates solely on negative feed-back principles, and it is of interest to enquire whether such simpler systems are found at the level of gene control. The discovery of such regulation was made at the same time, in fact, as the work which unveiled the principles of end product inhibition and lactose induction, the mid to late 1950s. A typical example is to be found in the control of the enzymes responsible for arginine biosynthesis, of which there are seven in *E. coli.* If arginine is present in adequate amounts in the growth medium, then these seven enzymes are not detectable in the culture. Upon removal of arginine, the enzymes begin to appear in cells within a period of five to ten minutes and accumulate to significant levels within an hour. This response time is the same as that for β-galactosidase induction and it was shown that the same basic considerations were satisfied: there was no gene selection, and new protein was being synthesized, so that the process involved was at the epigenetic level of control. It was also found that the synthesis of all the enzymes involved in the pathway is controlled simultaneously, as was the case with the enzymes of the *lac* operon. So these enzymes constitute a unit of epigenetic control. However, it was shown by mapping studies that the structural genes for the arginine enzymes in *E. coli* K12 do not map in one section of the chromosome, as do the *lac* genes. Rather they are scattered into three separate clusters, so that the co-ordinated control of their synthesis must occur by a pleiotropic effect of arginine; i.e. whatever the nature of the controlling signal, it must act simultaneously on similar control sites adjacent to each of the three structural gene clusters, each of which is presumed to behave in the manner of the *lac* operon, with an operator site adjacent to the structural genes as shown in Figure 2.2. Discovery of constitutive mutants with the same characteristics as those studied in relation to the *lac* system suggested that the same basic molecular machinery was at work, with one significant difference. Since the presence of a control signal (arginine) results in repression and its absence results in induction, it is evident that a postulated repressor of the arginine structural genes must be active only if arginine is present, and inactive in its absence. Thus instead of a repressor which binds to an operator site when free of ligand and is inactive if combined, as with the *lac* repressor in relation to lactose (or allo-lactose) as ligand, the arginine repressor must be unable to combine with operator sites unless it is combined with arginine. The evidence for this behaviour of the arginine system is given in papers by Maas (1961), Vogel (1961) and Gorini *et al.*, (1961). It was found in addition that arginine acted as a feed-back *inhibitor* of the first enzyme in this pathway, ornithine transcarbamylase (OTC).

We can now draw a control circuit for the arginine system, proceeding by analogy with *lac* operon but incorporating the features described above. This is shown in Figure 2.4. Here G_1, G_2, and G_3 are the three clusters of

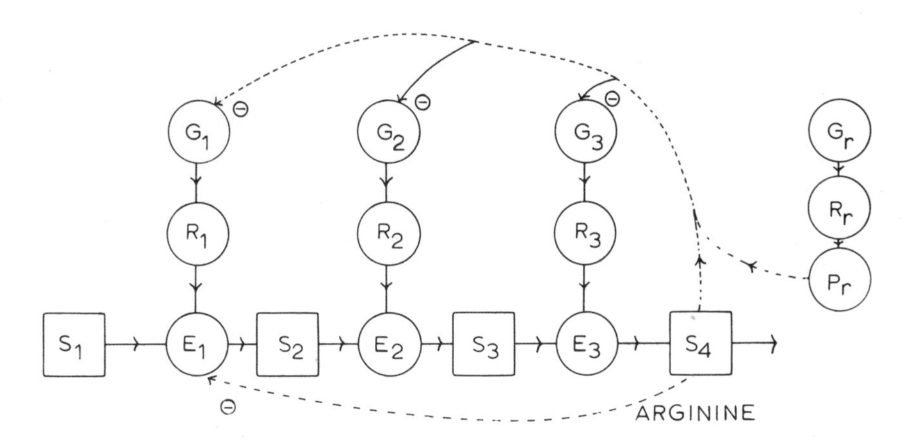

Figure 2.4. Feed-back inhibition and repression in the arginine control circuit of *Escherichia coli.*

structural genes and G_r is the regulator gene, producing the inactive repressor, P_r, referred to as an aporepressor. This aporepressor when combined with arginine as corepressor forms an active repressor which can interact with operator sites adjacent to the structural genes, preventing RNA polymerase action and so arresting transcription. As before the Rs refer to polysomes, the Es to enzymes and Ss to metabolites. There is some compression of notation in this Figure, since E_1 refers to the set of enzymes whose genes are clustered in G_1. Thus the feed-back inhibition effect of arginine is only to the first enzyme in the sequence, OTC, not to the whole set

We can now see one interesting distinction between feed-back inhibition and feed-back repression: the former acts on one enzyme only, the first in the pathway; the latter acts on all genes involved in the biosynthetic sequence. Again the logic is impeccable. Flow along an enzyme pathway can be controlled most efficiently through the first enzyme. Control of other enzymes as well would add nothing to this, and would in fact trap metabolites uselessly in pools from which they could not escape. But when a bacterium is offered arginine, it wants not simply to shut off its production at the enzyme level; it wants as well to shut off production of all mRNA and enzymes involved in its production, since their continued synthesis would be wasteful. Thus we see how the metabolic system and the epigenetic system work hand-in-glove to regulate vital end products, in this case, arginine, in relation to physiological supply and demand. Feed-back inhibition acts quickly (relaxation time about 10^2 s) and adjusts arginine levels in response to rapid fluctuations; feed-back repression acts much more slowly (relaxation time about 3×10^3 s) and adjusts macromolecular populations to appropriate levels for different nutrient conditions. Its more sluggish response is due to the slowness of macromolecular synthesis (about one molecule per five seconds) and degradation (a few minutes for mRNA) or dilution rates (of the order of an hour for protein). From Figure 2.4 we see that the hand-in-glove expression for the relationship between metabolic and epigenetic systems is actually more than a figure of speech. Evidently the metabolic system is contained in the epigenetic system, in the sense that whenever we are talking about enzyme adaptation, for example, we must at the same time be including the effects of feed-back inhibition, since the time periods for operation of the former necessarily include the latter. Thus in constructing a functional hierarchy for organismic physiology, the relationships between levels are inclusive, higher order systems containing lower order ones; but not of course the converse. These relationships can be formalized mathematically, as is done in my book *Temporal Organization in Cells* (Goodwin, 1963), but I shall make little use of such procedures in this volume.

It is evident that the arginine feed-back repression circuit is simpler than the *lac* operon system, in the same way that the feed-back control system of Figure 1.10, describing a biosynthetic sequence, is simpler than the glycolytic system described in Figure 1.11. There are undoubtedly good physiological reasons for these differences, the energy generating system to which the *lac* operon and glycolysis both belong perhaps requiring more rapid adaptive response, provided by the short positive feed-back circuits, than biosynthetic processes. However, there is a very considerable variety of control circuits now known, involving modifications of the principles so far discussed. An interesting discussion of different control strategies in relation to epigenetic and genetic stability is to be found in a paper by Savageau (1974). Also there are complex interactions between functionally-linked control circuits, such as those for arginine and for CTP, which are connected at their origins through carbamyl phosphate. Carbamyl phosphate synthetase appears to be under the joint control of pyrimidines and arginine (Gorini and Kalman, 1963; Pierard *et al.*, 1965), for example. Thus the view presented at the end of the last chapter of a cell partitioned into partially autonomous subunits of control, though still largely true at the next level of the epigenetic system, is complicated by an intricate pattern of interactions. Nevertheless it is important to realize that the basic principles of functional partitioning hold throughout the hierarchy of physiological levels, giving the system flexibility, adaptability and rapidity of response by allowing subroutines, as we may refer to them, to be called into operation or suppressed without extensive adjustments being made throughout the rest of the system.

Control of Ribosomal RNA Synthesis

The growth behaviour during enzymic adaptation as described in Figure 2.1 involves an interruption of exponential growth followed by resumption at a rate on the new carbon source (lactose) about equal to that on glucose. The general growth machinery of the cell remains intact during this adaptation, and only that part of the physiology directly related to the catabolism of lactose undergoes change. But what happens when the cells are transferred from medium which allows only relatively slow growth (say minimal salts with glucose, doubling time about 60 minutes) to nutrient medium which allows cells to divide every 20 minutes? What factors determine the change of overall growth rate?

A detailed investigation of this problem of the regulation of synthesis of the macromolecular species DNA, RNA and protein, was undertaken by O. Maaløe and his colleagues in Copenhagen on the bacteria *E. coli* and *Salmonella typhimurium*. Many of their findings are reported in the book entitled *Control of Macromolecular Synthesis* by Maaløe and Kjeldegaard (1966). An earlier paper concerning the essential observations is by Maaløe and Kurland (1963). They observed that during a change from a relatively poor to a rich medium (shift up) there is a dissociation between the behaviour of protein, RNA and DNA. The first of these to respond to the changed conditions is RNA, which increases in concentration within the cells rapidly and then levels out to a new rate some 15 minutes after the shift up. Only then does the rate of protein synthesis increase to the new rate and finally the DNA synthetic rate increases shortly after protein to the new value. All three must of course increase together at the same rate once a new condition of balanced growth has been reached. But the observation that RNA is the first to respond tells us two things: first that these macromolecular species can be dissociated from one another dynamically, so that they have distinct control circuits; and second, that RNA plays a crucial role in determining overall cell growth rate. Since the bulk of this RNA is ribosomal, with much smaller amounts of transfer RNA (tRNA) and messenger RNA (mRNA), we are dealing here with a control system for the ribosomal RNA genes. We have already seen how mRNA synthesis is controlled by mechanisms of repression.

Confirmation that ribosomal RNA (rRNA) is the rate-limiting factor for growth rate was obtained when it was shown by Maaløe and Kurland that the ratio of the rate of protein synthesis to the concentration of rRNA within a cell is constant over a wide range of growth rates. This relationship can be written in the form $\Delta P/R_r = \text{constant}$, where ΔP is the amount of protein formed per unit of time and R_r is the intracellular concentration of ribosomal RNA. An observational consequence of this relationship is that the faster cells grow, the more dense should the cytoplasm be with ribosomes, and this indeed is the case. A cell growing on nutrient medium with a doubling time of 20 minutes is packed with ribosomes throughout, except for the nuclear areas where the chromosomes are located, which are more thinly populated with ribosomes. Such cells are large, with an average of 3·1 nuclei per cell. Cells growing with a doubling time of 40 minutes are somewhat less than half this size, with about 1·5 nuclei per cell and about half the density of ribosomes; while cells with doubling times of 120 minutes are small, with a sparse population of ribosomes in their cytoplasm and about 1·2 nuclei per cell. We shall consider later why cells change their size with different growth rates.

In seeking a possible control mechanism which would explain the observation that ribosomal RNA synthesis can be transiently uncoupled from protein and DNA synthesis, but that in balanced growth there are well-defined relationships between these constituents, Maaløe and Kurland (1963) made use of two important bits of evidence. The first was that cells which require any of the 20 amino acids will not synthesize detectable amounts of rRNA, although they will do so if provided with the required amino acid even if protein synthesis is inhibited by chloramphenicol. The second observation was that if chloramphenicol is added to normal cells growing on glucose–minimal medium (i.e. without amino acid), there is a marked increase in the rate of rRNA synthesis similar to that occurring after a shift up, followed by deceleration of rate to a low value. The overall growth of such a chloramphenicol-blocked culture is of course arrested by the antibiotic. Together these observations point to amino acids as elements in the control link between growth conditions and rate of rRNA synthesis, since chloramphenicol blocks protein synthesis by preventing the amino-acyl-transfer RNA complex from transferring the amino acid to the ribosome.

If we follow the logic of the feed-back repression model, we should look for some metabolite which is a product of protein synthesis and could act as a co-factor in the repression of ribosomal RNA. Such a metabolite could be uncharged tRNA, i.e. tRNA free of amino acid. Cells which require any one of the 20 amino acids would have a population of such tRNAs corresponding to the absent amino acid species, and these molecules could then act as co-repressors, activating a protein repressor in the same way that arginine acts to repress the arginine structural genes. It would clearly be necessary for any one of the 20 different tRNA species to be capable of playing this role, but this is not an implausible requirement. With 70 to 80 nucleotides and a molecular weight of about 25 000 there could easily be a nucleotide sequence common to all tRNA species which is specific for interaction with a protein aporepressor, giving an active repressor. The chloramphenicol effect is then easily explained by the accumulation of amino acids in cells in which there is no protein synthesis, so that the tRNA is converted to the amino-acyl form and the rRNA genes are released from repression. The extent of this derepression would then be dependent upon the intracellular concentration of amino acids when the chloramphenicol block is imposed: if a cell has a high amino acid content already, the drug will cause little change in the concentration, and so no appreciable change in rRNA synthesis; but if the cell has a low level of amino acid, the chloramphenicol effect should be large. This is in accordance with observations on cultures growing under rich, medium and poor conditions.

From our previous considerations of feed-back repression circuits, it would be anticipated that it should be possible to find a constitutive mutant; in this case, one which makes rRNA at relatively high rates irrespective of the nutrient condition and hence the intracellular amino acid concentration. Such a mutant, referred to as 'relaxed' and given the letters *RC*, was reported by Stent and Brenner (1961). The wild type gene, RC^+, has the stringent control of rRNA synthesis described above in response to amino acid changes. The control picture these observations suggest is that shown in Figure 2.5. In this highly schematic drawing, G_r represents the ribosomal

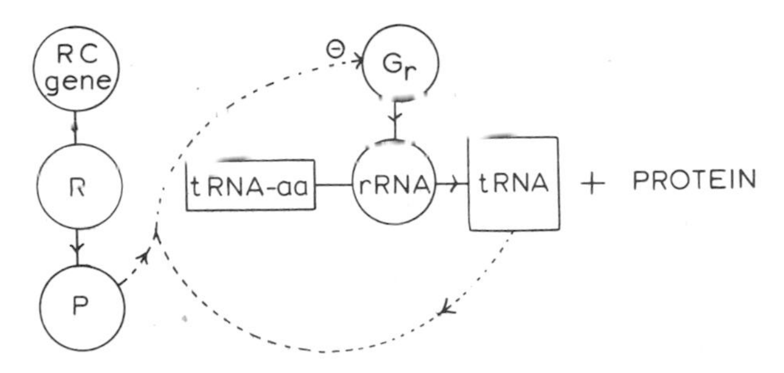

Figure 2.5. A schematic description of the circuit for rRNA control as initially proposed by Maaløe and Kurland (1963).

RNA genes and rRNA represents functional ribosomes, producing protein and tRNA from amino-acyl-tRNA precursors. Any of the tRNAs can function as a co-repressor with the aporepressor, *P*, product of *RC* gene activity. This control circuit operates on the same principles as that shown in Figure 2.4, and indeed it was designed on the basis of the feed-back repression prototype. Unfortunately, for all its elegance and plausibility, it fails to explain observations which have been made more recently and so requires modification and complication. The development of ideas about the molecular basis of rRNA control is in fact rather similar to that which took place with respect to catabolite repression, because before it was discovered that cyclic AMP was involved in the glucose effect, a somewhat simpler control circuit than that of Figure 2.3 had been proposed, based upon the principles of feed-back repression. Oncc again the evidence for this was the discovery, by Loomis and Magasamik in 1966, of a constitutive-type mutant which, like the RC^- mutant, showed 'relaxed' behaviour in that it failed to respond to glucose in the usual manner, making β-galactosidase in the presence of inducer (a necessary condition) irrespective of the presence of glucose. It was therefore proposed that this mutant, labelled CR^- (defective in catabolite repression), was making a deficient apo-repressor protein, which would normally combine with the catabolite repressor of Figure 2.3 and arrest transcription from the *lac* operon. This hypothesis is no longer acceptable. Although the full story about catabolite repression is far from being told, the current evidence suggests a mechanism somewhat like that of Figure 2.3, with the further complication that cAMP affects many other genes involved in catabolic processes as well as the *lac* genes. Constitutive behaviour of the type observed by Loomis and Magasanik could arise from a mutant in which E_u fails to respond to catabolite repressor. The picture that is now emerging regarding rRNA control via specific and somewhat unexpected metabolic species is rather similar to that deduced for the control of the *lac* operon, involving what may well be a high-level integrating control function by a newly discovered metabolite.

One important observation that cast some doubt on the validity of the control circuit shown in Figure 2.5 was that no effects of different amino acid conditions and different mutants on RNA polymerase activity could be detected. Another possible mechanism of action of the *RC* gene could be through the control of precursors of RNA synthesis, which control would therefore operate on *all* RNA synthesis, mRNA and tRNA as well as rRNA. It was suggested by Cashel and Gallant (1969) that a reaction normally involved in protein synthesis 'idles' when any species of tRNA is not charged with its cognate amino acid, and that this idling reaction, carried out by the *RC* gene product, generates an inhibitor of the phosphorylation reaction producing precursors for RNA synthesis. They showed by autoradiography of chromatographically-separated compounds that a metabolite does indeed accumulate in amino-acid starved, wild-type cells, but not in RC^- cells. This metabolite was a phosphorylated compound of unknown nature and so was appropriately called magic spot (MS) and given the label MS1. It was later shown to be guanosine tetraphosphate, ppGpp. A second phosphorylated substance was found at the same time and labelled MS2. It appears to be a detoxified form of MS1.

The discovery of ppGpp and its implication in the control of RNA synthesis stimulated a fairly intense study of the possible control circuitry of this system. The general picture, still incomplete in detail, which emerges from the work of investigators such as Peterson, Lane and Kalgora (1973), Hazeltine *et al.* (1973) and Yang *et al.* (1974) is as follows. The product of the *RC* gene is a protein of molecular weight about 80 000 which is called the stringent factor (SF). When this is included in a reaction system containing ribosomes, tRNA, mRNA, GDP and ATP, guanosine tetraphosphate is synthesized. It appears that SF is itself involved in the normal translation process, but in the absence of an amino acid the system produces ppGpp. This tetraphosphate is a strong inhibitor of the first enzymes in the biosynthetic pathways leading to GTP and ATP, so that a primary effect is to arrest the production of essential precursors for RNA synthesis. RNA polymerase is also somewhat inhibited by ppGpp and it decreases the half-life of rRNA (Gallant *et al.*, 1970). Thus the overall control picture is rather like that shown in Figure 2.6. Here G_r represents the rRNA genes and G_i any structural gene.

In the presence of amino-acyl-tRNA translation proceeds and little or no ppGpp is made. However, in the absence of an amino acid the ribosomal reaction complex produces ppGpp which then inhibits the production of GTP and ATP by inhibiting the first enzymes in the pathways from inosinic acid to the purine triphosphates, as well as inhibiting RNA polymerase.

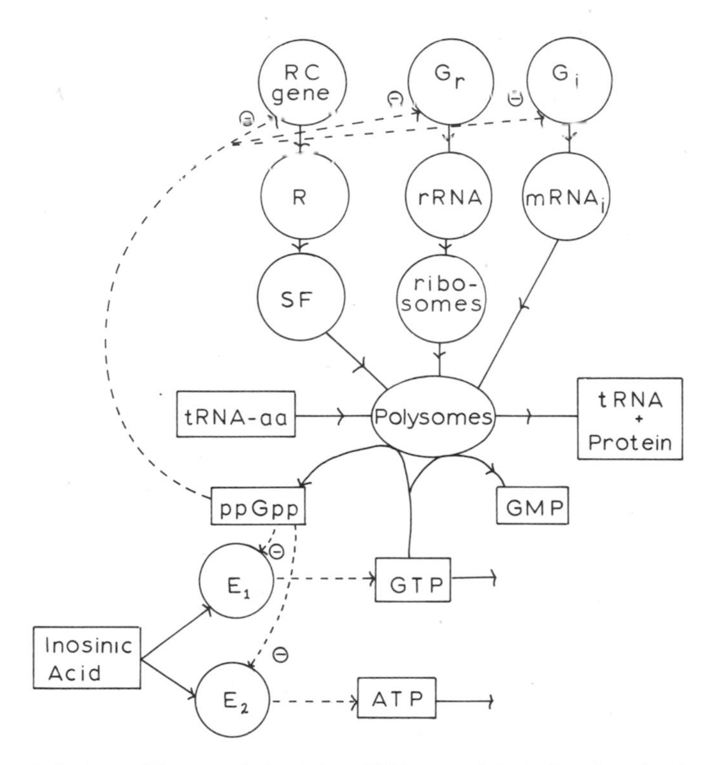

Figure 2.6. A possible control circuit for rRNA control, including the role of ppGpp.

Chloramphenicol appears to inhibit ppGpp synthesis directly, thus preventing the normal stringent response. The relaxed mutant, though capable of making protein, cannot make ppGpp and so fails to respond normally to amino acid deficiency. Thus we see that the rather simple picture of Figure 2.5 is made considerably more complicated. Instead of a feed-back repression loop specific for rRNA genes, we find a much more general feed-back loop involving a new metabolite playing a high-level control function in the system. The balance between the rates of protein and RNA synthesis is thus very finely tuned via the precursor pools for these macromolecules. Once again we see that a simple partitioning of the prokaryotic cell into functional units of control cannot be carried out because of the integrative effects arising from a co-ordinating signal. The prototype for unit control is the arginine biosynthetic pathway, where only arginine affects repression of the genes involved in the pathway, and these are the only genes affected (Gorini, 1963). However, the repression aspect of the *lac* operon and the control of RNA synthesis have shown us that more elaborate interactions are involved in the regulation of physiological function, and that co-ordination is achieved by special metabolites such as cAMP and ppGpp. These may usefully be regarded as the prokaryotic analogues of hormones.

Dynamics of Epigenetic Networks

I completed the last chapter with a section on the dynamics of metabolic control circuits, wherein it was shown that the occurrence of feed-back loops alters the stability characteristics of metabolic processes and complicates the dynamic possibilities very considerably. I will now do likewise for epigenetic control circuits, sketching some of the approaches that have been used in their analysis. The objective is to combine a study of the behaviour of the sub-systems which make up an entity with some analysis of the total system itself, which in this instance is the cell, functioning over epigenetic time intervals. It will again become quickly evident that there are severe limitations to this procedure, arising largely from the difficulties of non-linear, global analysis.

Analogy with Statistical Thermodynamics

One possible approach to the dynamics of cell behaviour is to proceed on the analogy between the cell seen as a collection of interacting control circuits and a gas seen as a collection of interacting molecules. At first sight this may seem to be a very unpromising analogy. The aggregate properties of a gas such as its temperature, pressure, volume and compressibility, can be calculated from the properties of individual molecules (mass, velocity, degrees of freedom) and their interactions by averaging operations, hence the expression statistical thermodynamics. These averaging operations depend upon the fact that individual molecules behave independently of one another, so that there is no co-operative or collective behaviour. There can, however, be constraints arising from interactions such as van der Waals, electromagnetic or other forces which can give collective effects, and the methods of statistical thermodynamics can also be used for liquids and crystals, where more severe constraints operate. Such investigations are of considerable importance and interest, and phenomena such as discontinuities and phase transitions can be explored and analysed. In the last chapter I shall apply some of these methods to such phenomena as cell growth and membrane excitability, where the consequences of co-operative interactions will be studied. However, it is of some interest now to look at the picture which emerges when one views the cell as a collection of functional units, governed by particular control dynamics, when these are in weak interaction with one another and no collective behaviour occurs.

I explored this treatment of cell behaviour in some detail in my book *Temporal Organization in Cells.* My Ph.D. thesis, written in 1959 and from which the book originated, was based upon a view of gene regulation antedating the theory of feed-back repression which emerged in the early 1960s. This was the plasmagene theory as developed particularly by Spiegelman and Reiner (1947) and Spiegelman (1948) in relation to enzyme adaptation in yeast. Monod (1947) in his review refers to this theory and discusses a modified form of it. Its essence was that the regulation of gene activity in cells arose primarily from a balance between competing autocatalytic or self-replicating units whose relative proportions could be altered by environmental factors such as substrates. This is clearly an evolutionary-type model of intracellular processes, where there are natural selection forces operating on self-reproducing and competing units. Before ideas about molecular feed-back control processes became operative in cell biology in the late fifties and early sixties, models of cell differentiation based upon the evolution paradigm were very common, and in fact go back to Wilhelm Roux. The plasmagene theory was one of these, being in fact more an ecosystem model than an evolutionary model, since selection was by different environments for species which already existed, there being no analogue of gene mutation. The self-replicating or autocatalytic enzyme forming systems were the units out of which I constructed a statistical thermodynamics of cell behaviour, with predator-prey types of interaction between these units giving oscillatory dynamics to the whole cell, and a certain type of discontinuity in relation to a 'temperature' function which I used to explain cell differentiation. However, when I became aware of the evidence for specific feed-back control processes as the basis of gene regulation in prokaryotes, I used a physiological unit of the type shown in Figure 2.4 as the basis for the analysis. Functionally such a unit can be telescoped into the simpler form shown in Figure 2.7, assuming that one of the elements of the feed-back repression system is rate-limiting for the behaviour of the whole, and that this one only is being represented. The number of steps in the enzyme sequence will depend upon which loop is taken to be rate-limiting. Very crude equations for transcription and translation may be derived on the assumption that mRNA (X) is rate-limiting for protein synthesis and enzyme (Y) limits the rate of production of substrate, S. Also it is assumed that S limits the rate of message transcription at G. These assumptions are reasonable in view of the observation that if some other factor were rate-limiting (e.g., amino acids or protein synthesis) then the feed-back repression circuit shown would not effectively be operating.

Using an expression of the type in equation (1.10) for the repression effect of S on the gene, we can write

$$\frac{\mathrm{d}X}{\mathrm{d}t}=\frac{a}{b+cS^n}-dX \tag{2.1}$$

where a and b contain all the terms for substrates, rates and the enzyme

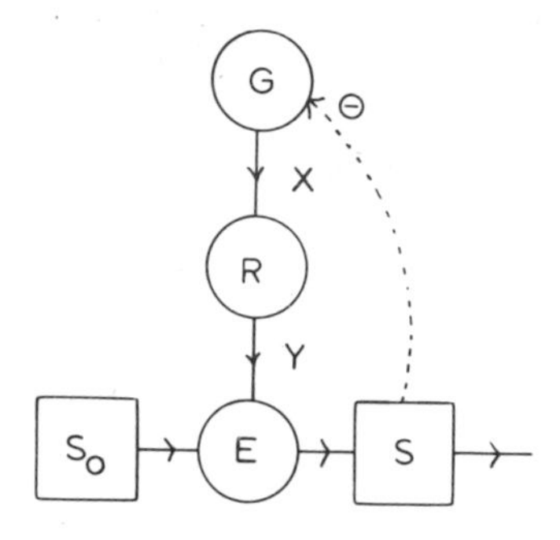

Figure 2.7. A simplified representation of feedback repression, where G = gene, R = polysomes, E = enzymes, S_0 = substrate, and S = product which feeds back to the gene and represses it.

involved in RNA synthesis; c is an affinity constant of the repressor (n molecules of S complexed with apo-repressor) for the gene; and d is the rate constant for degradation of mRNA. For protein synthesis, we can use the simple equation

$$\frac{\mathrm{d}Y}{\mathrm{d}t}=eX-fY \tag{2.2}$$

where e contains all the rate constants, substrates and factors involved in protein synthesis, and f is the rate of protein degradation. Finally, assuming that substrate concentrations are small so that the Michaelis expressions are linear and S_0 is constant, we can write

$$\frac{\mathrm{d}S}{\mathrm{d}t}=gY-hS. \tag{2.3}$$

These equations are qualitatively similar to those derived in the last chapter, equation (1.25), with the simplification that we have linear terms for all but the control expression in equation (2.1). Thus an analysis of their stability characteristics can proceed by the same technique as used there, linearization and study of the characteristic roots. Once again the studies of Walter (1969, 1970), Viniegra–Gonzalez and Martinez (1969) and Rapp

(1975), are of value in telling us what the relationship must be between the number of steps in the reaction sequence and the stoichiometry of the control reaction (value of n) to give stability or instability (oscillatory behaviour).

In order to pursue the analogy with statistical mechanics, it was necessary to modify equations (2.1), (2.2) and (2.3), so that they could be integrated to give a first integral of the motion; i.e. they had to be put into conservative form. The terms dX and fY in equations (2.1) and (2.2) respectively, were replaced by constants, d and f. This means that instead of degradation rates being proportional to the concentrations of the molecules being degraded, it was assumed that mRNA and protein were degraded at fixed rates per unit time. This would correspond to the special situation in which the ribonucleases and the proteases are always present in rate-limiting concentrations, and that mRNA and protein are inactivated only by catalytic degradation. Clearly these are restricting assumptions, but we shall see later that the restrictions are not so severe as might be expected.

The equations can be further reduced by using the general argument already introduced about the differences in the characteristic response times of the epigenetic and metabolic systems. Equation (2.3) refers to a metabolic reaction in the variable S. This will proceed very quickly in relation to the time required for a significant change in Y, so we can take it that S will always be approximately at its steady state and write

$$0=\frac{\mathrm{d}S}{\mathrm{d}t}=gY-hS$$

Therefore $S=(g/h)Y$, and we can then eliminate S from equation (2.1). To keep the argument at its simplest, let us take $n=1$, so that we get equations of the form

$$\left.\begin{aligned}\frac{\mathrm{d}X}{\mathrm{d}t}&=\frac{a}{b+kY}-d\\ \frac{\mathrm{d}Y}{\mathrm{d}t}&=eX-f.\end{aligned}\right\} \qquad (2.4)$$

These can evidently now be integrated, as follows

$$(eX-f)\frac{\mathrm{d}X}{\mathrm{d}t}+\left(d-\frac{a}{b+kY}\right)\frac{\mathrm{d}Y}{\mathrm{d}t}=0$$

$$e\frac{X^2}{2}-fX+dY-\frac{a}{k}\log(b+kY)=\text{const.},$$

the constant being determined by the initial conditions for the variables X and Y.

Now that we have an integral, it is necessary first to find out what it signifies. This is most easily done by first transforming equation (2.4) to new variables, defined as follows:

$$x = X - \bar{X}$$

$$y = Y - \bar{Y}$$

where $\bar{X}$ and $\bar{Y}$ are the steady state values of the variables obtained by setting $\mathrm{d}X/\mathrm{d}t = 0$, $\mathrm{d}Y/\mathrm{d}t = 0$ in equation (2.4).
The equations then take the form

$$\frac{\mathrm{d}x}{\mathrm{d}t} = d\left(\frac{\gamma}{\gamma + y} - 1\right)$$
$$\frac{\mathrm{d}y}{\mathrm{d}t} = ex \qquad (2.5)$$

where $\gamma = a/dk$.
These equations give the integral

$$\frac{ex^2}{2} + d[y - \gamma \log(\gamma + y)] = C. \qquad (2.6)$$

If C is greater than 0, the expression on the left-hand side describes a closed curve in the xy-plane. This is clear from the fact that any line $y = k$, constant, will cut the curve in two points, the solutions of $ex^2/2 = C - d[y - \gamma \log(\gamma + y)]$, which will be real if the right-hand side is non-negative. It can also easily be shown that any line $x = k$, constant, will cut the curve in two points, the solutions of

$$y - \gamma \log(\gamma + y) = C - \frac{ek^2}{2},$$

providing again the right-hand side is non-negative. The shape of the curve is as shown in Figure 2.8. This egg shape reflects the non-linearity of the control term in equation (2.5) and always seemed to me a very promising foundation for a biological theory. Unfortunately it has not yet hatched.

Evidently, then, what equation (2.5) generates is continuous oscillations, the variables undergoing continuous cycles about their steady-state values. Therefore any construction of statistical mechanical type based upon such equations could tell us something about on-going rhythms in cells, arising from oscillations in feed-back repression circuits. From our estimates of epigenetic relaxation times, we might expect the periods of such oscillations to be a few hours. What behaviour of cells could such rhythms underlie? One obvious suggestion is the cell cycle itself, although in order to model this properly, modifications of the equations would need to be introduced. Another possibility is connected with the phenomenon of biological clocks, where continuous oscillations occur with periods in the neighbourhood of 24

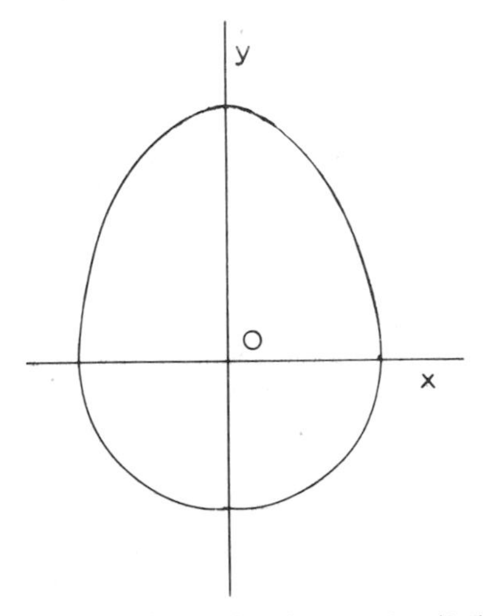

Figure 2.8. The closed curve of y against x given by equation (2.6).

hours. Both of these are epigenetic phenomena as defined in this chapter and they could be analysed by the procedures described. To construct a statistical mechanics for equations like (2.5) and others of a more complex nature related to these, one proceeds from integrals of type equation (2.6), one for each control circuit regarded as a contribution to the overall dynamics and from these one constructs a density distribution function which describes the probability of occurrence of any particular possible state of the whole dynamical system consisting of all the control circuits and their interactions, however these have been defined. One can then proceed to ask questions about various properties either of the whole or of the parts. The operation of embedding a part in a complex whole of this kind allows one to use some fairly powerful analytical procedures and to get statistical answers to problems which would be almost insoluble if one were to ask them of the deterministic system from which the analysis started, namely equations like (2.5).

One might at this point return to the derivation of equation (2.5) and object that the assumptions made in getting these are so restrictive as to invalidate their use *ab inito*. It is for example easy to show that the original equations (2.1)–(2.3) cannot oscillate if $n = 1$ and that oscillations arise only because the mass action degradation terms have been left out. However, it is generally recognized that if stochastic features and time lags are introduced into deterministic equations, such as allowing the parameters a and e to undergo random fluctuations, then an otherwise stable system can become unstable and start to show oscillatory behaviour. Tiwari and Fraser (1973) have recently shown that this occurs with equations (2.1)–(2.3) when the inevitable noisiness and random delays of intracellular molecular events are allowed for by introducing stochastic variability into the process of mRNA passage from gene to ribosome, the translation process and macromolecular degradation. Thus equation (2.5), when embedded in the probability ensemble of statistical mechanics, may give a fairly accurate picture of intracellular processes.

The other criticism, however, is more telling. It is that such procedures lead essentially to averaging operations and these are of limited value in relation to the major preoccupations of current biological enquiry, which are primarily concerned with order arising from order, not order from disorder as Schrodinger put it in his still very interesting and provocative monograph, *What is Life*? (1948). A basic contention in that book was that statistical mechanics is an inappropriate model for biological process because organismic order does not depend upon averaging over large numbers of elementary events, but upon dynamical coupling of a particular kind which effectively avoids the dissipation of energy in the form of heat or thermal noise. The analysis of the behaviour of such dynamically ordered systems is not, however, inconsistent with the postulates of statistical mechanics and studies of coupled, organized, biological systems within this context have been made by Markowitz (1971), McClare (1971), Kornacker (1972) and Markowitz and Nisbet (1973). A rather general approach to the study of biological order from a thermodynamic viewpoint has been advanced by Glansdorff and Prigogine (1971), using a concept which they call a dissipative structure, one which utilizes or dissipates energy to maintain its order. This is, of course, the open system, made familiar to biologists by von Bertalanffy (1952) and others more than two decades ago. What Prigogine and his colleagues have done is to demonstrate certain necessary thermodynamic conditions for the existence of particular classes of dynamic behaviour in far-from-equilibrium systems, but what is missing from their analysis is a thermodynamic or macroscopic treatment of these dissipative structures. Only rather recently have analytical methods for the phenomenological description of dynamically ordered steady states of open systems been developed, some of which are still very controversial. I shall consider their application to certain biological processes in the last chapter.

Analogy with Computers: the Cell as a Finite Automaton

What appears at first sight to be a totally different approach to cellular behaviour, seen as a result of interacting epigenetic regulatory circuits, is based upon the view of genetic control processes as logical switching circuits. The suggestion that gene activity could be looked at in terms of such binary on–off elements appeared in a number of publications such as those of Monod and Jacob (1961), Sugita (1963) and Apter (1966). However, a systematic analysis of the behaviour of large ensembles of such units was first carried out by Kauffman (1969), with some very interesting results. Kauffman's assumptions in constructing his simulation of the epigenetic behaviour in a network of functionally interacting genes were the following:

1. A gene is either *on* or *off* and so can be assigned the value of 1 or 0, respectively.
2. Each gene receives only two 'inputs', i.e. two metabolic signals (e.g. the *lac* operon responds to the repressor and to cAMP in Figure 2.3). This assumption can be weakened to 'a small number of inputs' without substantially altering the results.
3. The inputs can be positive or negative (again like the *lac* operon, where repressor has an inhibitory and cAMP a stimulatory effect) and assignment of inputs and connections to genes is according to random selection. Thus the networks studied are randomly constructed, with roughly equal proportions of all possible combinations of inputs and connections.

Using these basic postulates, the behaviour of a gene can be represented by a state transition table as follows. Suppose we are considering gene X with inputs from genes Y and Z. These inputs are of course not directly from these genes, but occur via gene products, as repressor is the product of the i gene and cAMP the product of the gene coding for adenyl cyclase. Time is quantified in unit steps in such a logical machine or automaton, so we can represent the state transition of gene X as shown in Figure 2.9. The pairs of possible input values to the gene at time T are shown under the input variables, Y and Z, while the output of X in response of these inputs is shown at time $T+1$ under the variable X. The Figure is constructed for the *lac* operon with $Y = i$ gene and $Z =$ gene for adenyl cyclase. The *lac* operon is active only if the repressor is inactivated ($Y = 0$) and cyclic AMP is present ($Z = 1$). This table defines what is known as a Boolean function of the binary variables Y and Z:

$$X = f(Y, Z)$$

For logical functions of this kind, one uses tables instead of graphs.

Evidently there are many possible Boolean functions of two variables, obtained by placing 0's and 1's in different combinations in the column under X. Since there are four possible places and a choice of two variables for each, the total number of Boolean functions of two variables is $2\times2\times2\times2=2^4=16$.

T		T+1
Y	Z	X
0	0	0
0	1	1
1	0	0
1	1	0

Figure 2.9. The Boolean function which is taken to describe the switching behaviour of the lac structural genes in response to the inputs deriving from the activity of the *i* gene (Y) and the gene controlling adenyl cyclase production (Z). X represents the activity of the structural genes.

A network of interacting genes is constructed by deciding on some total number of genes, say $N=100$, and assigning to each one, by random choice out of the 16 possibilities, a particular Boolean function. Inputs to the genes are then assigned by random choice from among the total number of genes, one gene being allowed to function as input to several others if it is selected several times. With N genes in a net, the state of a net at any time will be given by a vector consisting of a series of N 0's and 1's, in the order of numbering of the genes. There will evidently be 2^N possible states for a net. Starting from some initial condition, say all genes off (0), the net will then move successively through a series of determined states according to the transition rules (Boolean functions) assigned to each gene. It was the behaviour of many such nets which Kauffman studied, using a computer.

One might have anticipated that nothing of obvious significance would emerge from such a study. For $N=100$, 2^N is a very large number. These networks have therefore many states available to them, and might be expected to meander through these states with no evident order. Since they are finite automata, they must cycle with a finite period, but this could be very large, with cycle lengths of 2^{100} or so. This is in fact the result one gets if the number of inputs per gene is N, the number of genes. As Kauffman points out, if a state transition requires one microsecond, then the time for the net to traverse a cycle is about $10^7\times$Hubbel's estimated age of the universe. No biological system has that much time available to it.

Networks with one input per unit have been shown to be little better for mean cycle times. It was therefore somewhat startling when Kauffman found that with connectivity 2 ($K=2$), short cycle times were obtained, the mean value being of order $\sqrt{N}$. So for $N=100$, the average cycle time for a net is ten units. If each unit is equal to one minute, then the mean cycle time is ten minutes.

At this point it is necessary to consider the biological system again, and attempt to make some estimates for N and the transition time. N is the number of genes in a cell and the transition time is the time required to activate a gene and get a significant metabolic response. We have estimated the latter to be several minutes to one hour. For unit responses of the type considered we may take this time to be about ten minutes.

Estimating the number of genes in a cell is a much less reliable business. There are of the order of 10^3 genes in bacteria and 10^4–10^5 in higher organisms. Let us take $N=10^4$. Then with one time unit equal to ten minutes (Kauffman's estimate is one minute, but this is rather too small for metazoan cells), average cycle times are $10\sqrt{10^4}=10^3$ minutes or $16^{2/3}$ hours. This is in the same time range as that estimated for oscillations in the epigenetic system, and falls within the domain of metazoan cell cycles. Kauffman applies his arguments to cell cycle times in different cell types, relating these to the total DNA content. Evidently one is on very shaky ground here, and I do not believe the details are important. What is interesting is the remarkable convergence of the two approaches on the phenomenon of rhythmic activity in the epigenetic system and the expected range of period. The suggestion is that intracellular dynamics of gene control processes is quite likely to involve oscillatory activity and that rhythmic behaviour of cells such as the cell cycle and biological clocks seem to be likely consequences of this. An interesting feature of Kauffman's analysis is that the system need not be designed with any foresight; most random nets will show stable cycling behaviour with reasonable cycle times. This may tell us something very significant about order from randomness which is of considerable evolutionary significance, thus introducing statistical design considerations which complement in an interesting manner the statistical dynamical arguments considered earlier.

While Kauffman's approach to epigenetic processes is robust in relation to generality of construction, it is weak in relation to the possibilities of mathematical analysis. This is exactly the converse of the statistical dynamical approach sketched above, where the analysis is possible but at the expense of generality of interactions between the units. However, more recent studies by Glass and Kauffman (1973) have shown how to introduce certain forms of analysis into this model; while Kauffman (1973) has extended his own original ideas in interesting ways. In applying the idea of switching circuits to the changes of epigenetic state associated with the phenomena of determination and transdetermination in *Drosophila*

(Hadorn, 1966), some very suggestive conclusions regarding spatial boundaries in the developing insect embryo are derived from considerations of state transition probabilities in *Drosophila* imaginal discs. That such deductions can be drawn from a model of epigenetic control processes based upon the simple foundation of Boolean logic is indicative of the unexpected power of this approach, as well as the imagination and clarity of Kauffman's work.

Chapter 3

THE MITOTIC AND CELL CYCLES

Prokaryotes

In the metabolic system of the cell as defined in Chapter 2, enzyme concentrations entered equations as parameters, remaining constant over metabolic relaxation times. In the epigenetic system, enzymes, mRNA and rRNA became variables as well as metabolites, while gene 'concentrations' were taken to be constant, defining some of the parameters. At the next level of our analysis, gene concentrations are allowed to vary and the control of DNA replication and associated events of the mitotic and cell cycles become the major focus of attention. Since there is an extensive overlap between the time periods of the mitotic cycle and epigenetic phenomena as described in Chapter 2, it is not reasonable to distinguish between them on the basis of characteristic relaxation times. Furthermore, we shall see that control of DNA replication appears to operate in accordance with essentially the same principles as regulation of gene activity, which is not surprising since DNA is just another macromolecular species. Therefore we shall include the mitotic and the cell cycles within the category of epigenetic control processes. The constraint operating here is that the genetic information store remains unchanged: i.e. mutations and processes such as crossing over, inversions, duplications, etc., which could affect genetic information and its accessibility, are excluded. These processes provide the source of variation at the next level of functional analysis of cells and organisms, defining the genetic system. I will not consider these processes in any detail, because at the moment they appear to be governed by principles quite different from those operating in regulatory activities; i.e. the current convention is to regard changes in genetic information as arising from purely random events over which cells and organisms have little or no control.

I start the analysis with the prokaryotic cell cycle in view of its comparative simplicity and the existence of sufficiently detailed information to provide a basis for molecular model building. Since a cycle has no beginning nor ending, a reference point must be chosen according to some functional criterion which hopefully reflects a significant feature of the intracellular process. There are two such reference points in the bacterial cycle, the first being the moment of separation of the two daughter cells and the second the moment when replication of DNA is initiated. Although the former is the more obvious one to choose for observational purposes, the latter turns out to be functionally more satisfactory. There is considerably less variation in DNA initiation than there is in cell separation, indicating that cells are more concerned with its accuracy.

Up to now no mention has been made of spatial organization and cytological detail, except for the briefest mention of the membrane boundary in relation to the behaviour of the *lac* operon. It is characteristic of a control theoretic approach to cell behaviour that time figures importantly in relation to dynamics, but space hardly at all. The cell might as well be a homogeneous sphere, bounded by a membrane, with all the constituents, DNA, RNA, protein and metabolites dispersed uniformly within this sphere. But the moment one starts to deal with the cell cycle, it becomes necessary to take account of some spatial constraints, although these are minimal as yet. They will become more important later on in relation to developmental processes, when field equations incorporating spatial variables will have to be used in place of ordinary differential equations. For the moment we need only observe that in a population of bacteria growing under uniform, non-crowded conditions, the cells have a very uniform size at any reference point in the cycle, indicating that they are able in some sense to measure their own space and to separate it into fairly equal halves at division. The bacterial species that have been used primarily for the analysis of the cell cycle, *E. coli* and *Salmonella typhimurium*, are roughly cylindrical in shape and grow along the long axis. The first evidence of a partition in the cylinder is a ridge, called the septum, which runs circumferentially around the cylinder in the middle of the long axis. The placing of this septum is evidently under accurate control, so that the cell has some spatial information which it uses to locate this structure (Walker and Pardee, 1968).

The normal temporal relationships occurring between DNA replication and cell separation can be altered by a variety of factors, showing that these events are not causally linked in any obligatory manner. For example, poor nutrition, Mg^{2+} starvation, penicillin, crystal violet, u.v.- and X-rays all prevent septum formation and so arrest cell division, but DNA replication and general growth continue. The result is a very long cell referred to as a filament. DNA replication can also be dissociated from growth by mitomycin C, phenethyl alcohol, or thymine starvation, so cells grow large without making DNA (Pardee, 1968). It is even possible to find mutants in which separation and cell division continue in the absence of DNA replication, so that cells are generated without any DNA. Such cells cannot, of course, survive very long (Alder *et al.*, 1967; Inoye, 1969). These dissociations demonstrate that the periodic events of initiation of DNA replication and initiation of septum formation are capable of autonomous cycling, so they have their own control mechanisms or else are both driven by another oscillator.

Since the bacterial chromosome is a closed circular loop of DNA, of circumference about 1·3 m, it would seem reasonable to suppose that a clock could be generated most simply by having the replication process run round and round the chromosome at a rate proportional to the growth rate of the bacterium. However, the evidence is decisively against this (Yoshikawa, 1967). Rapidly-growing bacteria increase the overall rate of DNA replication not by increasing the rate of addition of bases at the growing points, but by increasing the number of growing points: i.e., initiating a second round of replication before the first one is complete. Thus a clock must be in the nature of an oscillator controlling initiation events by cyclic concentration changes rather than an automaton travelling around a track. It then seems reasonable to suppose that a similar oscillator controls septum initiation and that these two oscillators are normally coupled in a particular phase relation. This relation is that septum formation begins when DNA replication ceases, there being then a period of about 20 minutes referred to as D (Helmstetter and Cooper, 1968) before cell separation occurs. The normal period of time required for completion of chromosome replication, C, is about 40 minutes irrespective of the growth rate (Helmstetter and Pierucci, 1968). These relationships define the moment of initiation of DNA replication in a cell growing with any particular growth rate. For example, if the doubling time is 30 minutes, then DNA replication terminates 20 minutes before division and initiation was 40 minutes before this. So the initiation of replication of chromosomes partitioned between daughter cells coincided with the division giving rise to their grandparents. Evidently initiations must occur in cells of such a culture every 30 minutes. If the generation time is altered to 50 minutes, then initiation of DNA replication for daughter cells occurs 60 minutes before their separation, i.e. 10 minutes before the division giving rise to their parent cell. Thus we see that DNA replication can be initiated at any time in the cell cycle, but this time will be fixed for any given generation time.

These structural and temporal relationships are shown diagrammatically in Figure 3.1 for a 50-minute cell cycle. Initiation of DNA replication occurs at zero time (same state as at 50 minutes) and is shown by the occurrence of a second dot on the membrane adjacent to the attachment site of each chromosome. This site defines the chromosome origin, DNA replication being initiated here and proceeding in both directions until the two replication forks reach the terminus, a point approximately opposite the origin (Masters and Broda, 1971). As replication and growth proceed, these sites move apart by the growth of new membrane between them, thus carrying the daughter chromosomes apart. The evidence for this was first presented by Jacob, Ryter and Cuzin (1966) in a study of *Bacillus subtilis*. Membrane growth in prokaryotes thus plays the role of the mitotic apparatus in eukaryotes, separating daughter chromosomes. The chromosomes are of

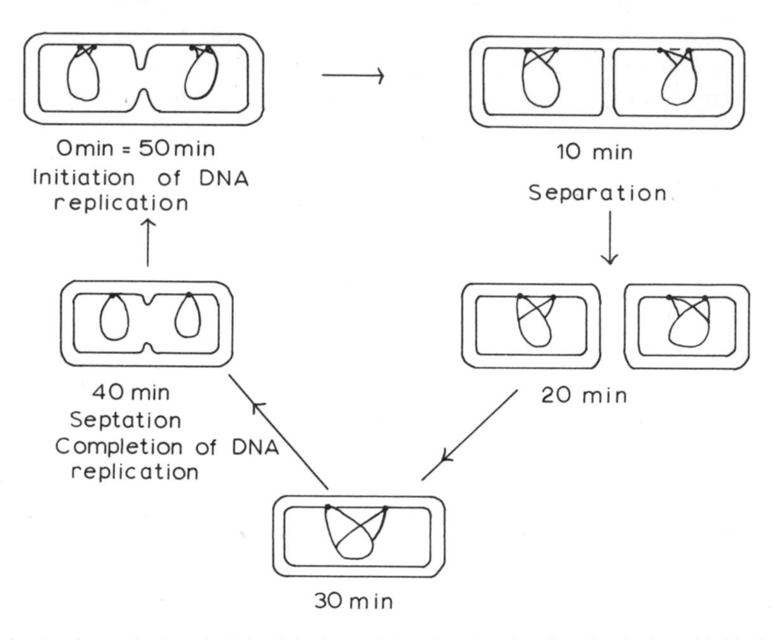

Fig. 3.1. A schematic description of the bacterial cell cycle, showing the timing of initiation of DNA replication, of septum initiation and of cell division in a 50-minute cycle.

course double helixes and replication is semi-conservative, but for simplicity the chromosomes are shown as single threads. Termination of chromosome replication coincides with initiation of septum formation. A very lucid and complete discussion of these relationships is given in a review paper by Donachie, Jones and Teather (1973).

The Replicon and the Initiator Hypothesis

One of the first molecular hypotheses to be advanced to explain control of DNA replication was based naturally enough upon the operator–repressor model of gene control discussed in the last chapter. The evidence on which it was based is the following. It had been observed that the capacity for autonomous replication does not belong to any fragment of bacterial DNA introduced into a cell, but only to a functionally integral unit, referred to as a replicon. The bacterial chromosome is one such unit, but there are others such as the sex factor of *E. coli* and the temperate bacteriophages. Each of these contains the information required for its own replication and so constitutes a unit of control. Mutants should then exist which have defective control in various ways. Jacob, Brenner and Cuzin (1963) reported on the occurrence of such mutants and introduced the replicon hypothesis. They discovered that two genetically distinct mutants both caused the cessation of DNA initiation. One of these was postulated to be responsible for the production of a protein required for initiation of DNA synthesis and was called the initiator gene. The other was assumed to be concerned with the attachment of DNA to the cell membrane, a necessary condition for replication. It was called the replicator gene and was assumed to define the origin of the chromosome. This control circuit then operates by positive feed-back, it being assumed that once in every replication cycle the initiator and replicator genes become active, resulting in the synthesis of the proteins required for another round of initiation. It is obviously an incomplete hypothesis, because it does not say what controls the timing of the activity of these genes, nor how their frequency of activity is related to the cell generation times. A model describing how this could occur has been advanced by Sompayrac and Maaløe (1973).

The influx of physicists into biology during the late 1950s and early 60s had led to the introduction of ideas about 'elementary particles' of cell function, given neologisms with the suffix 'on' to conform to physical usage as in the terms photon, electron, positron, etc. The naming of new units such as the muton, the cistron, the zymon became a fairly popular sport in molecular biology and the replicon added to these. A letter in the *Journal of Theoretical Biology* not long after the christening of the replicon suggested that only one more unit was needed to complete our understanding of cell organization: the leprechon. With it, all aspects of cell behaviour would be fully understood. It would seem from current research that, unlike the prokaryotes which have only one leprechon each, eukaryotic cells must have many such units, so that the concept not only completes the picture for bacteria but at the same time provides a full explanation for the bewildering behaviour of eukaryotic regulatory processes.

The Initiation Repressor Hypothesis

The replicon hypothesis depends upon the observation that particular proteins are necessary for the initiation of DNA replication. It is hardly surprising that such proteins should exist, but this in no sense establishes them as normal intracellular controllers of initiation. They may simply function as necessary conditions for initiation, in the same sense that an enzyme is a necessary condition for the reaction it catalyses to proceed at detectable velocity within cells. We have seen that enzymes are not themselves the control signals in the metabolic or the epigenetic system. Therefore it is perfectly consistent with the occurrence of initiator and replicator mutants that the final control of initiation of DNA replication be resident in the action of another protein, acting as an initiation repressor. Such a model would be more in line with the original repressor–operator model than the initiator hypothesis. We shall now see that it can give us also an idea of one way in which a cell can measure its own volume.

Pritchard, Barth and Collins (1969) suggested that each replicon contains a gene which codes for a protein with the capacity to prevent initiation of DNA replication either by interaction with initiator or with the chromosome origin, hence acting as a repressor of initiation. They supposed such a gene to be located just after the initiation site (the postulated replicator gene) on the DNA, which is located at a fixed point on the chromosome for any given strain of bacterium. Soon after DNA replication has begun, this repressor gene will be duplicated and, it is supposed, there will be a burst of mRNA produced from it. This then gets translated into repressor protein, H, which prevents another round of DNA replication from being initiated. Replication then proceeds at its normal rate, and the cell will grow at a rate dependent upon nutrient conditions. The concentration of initiation repressor will decrease as the volume of the cell grows, since it is assumed that no further synthesis takes place after the burst following upon duplication of the gene, and that the protein is stable. Another round of DNA replication can be initiated only when the concentration of repressor has fallen below some threshold value, which will occur when the cell has reached a certain size. Thus we see how increase of cell size connects causally with control of initiation and how a cell can 'measure' its own size. The concentration of the repressor within a cell will then rise and fall in a saw tooth manner as shown in Figure 3.2, initiation occurring at the trough of the oscillation. The exact shape of the descending part of the curve depends upon whether the cell

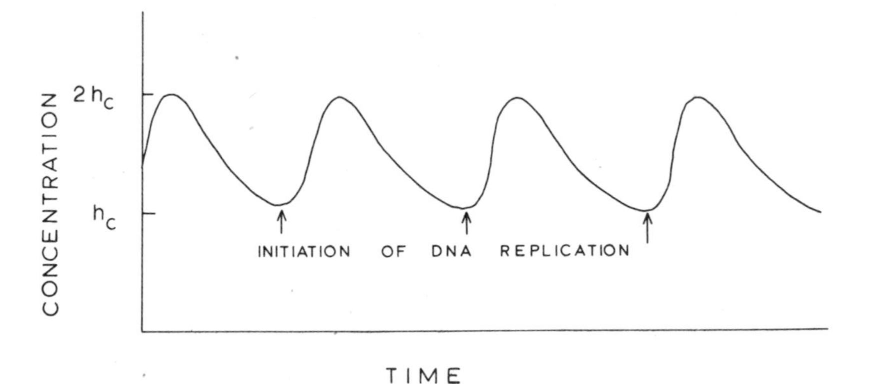

Figure 3.2. The hypothetical oscillator underlying the control of initiation of DNA replication in the model of Pritchard *et al.* (1969).

growth is linear or exponential, the latter alternative being shown in this instance. Cell division can occur at any defined point on the cycle, determined by the growth rate, as discussed above. Notice that cell division does not alter intracellular concentrations and so has no affect on a curve such as that shown in Figure 3.2. The amplitude of the oscillation is $2h_c$ where h_c is the threshold for initiation, since the factor of 2 in cell volume change necessitates the same factor in concentration change, according to the above postulates. There are several consequences of this model which are pursued in the article by Pritchard *et al.* (1969). One of these is the relationship between cell volume and generation time, g, which arises quite naturally from their hypothesis. The possibility of obtaining such a relationship from the constancy of $C+D$, the chromosome replication time plus the time for septum formation, as established by Helmstetter and Cooper (1968), was first pointed out by W. D. Donachie. He has since presented evidence for cell size regulation in terms of the concept of a unit cell (Donachie and Begg, 1970; Donachie *et al.*, 1973).

Let us choose some reference volume for a bacterial cell and call it V_r. A convenient volume to choose is that corresponding to a cell which is growing with a generation time of $C+D=60$ minutes, since such a cell will have one pair of replication forks initiated at each cell division. Let us next assume that the growth of single cells is exponential, so that we can write

$$V=V_r 2^{t/g}$$

where g is the generation time of the cell. Evidently for the particular case $g=60$, it is clear that at $t=60$ minutes $V=2V_r$, and the cell will divide. However, suppose that the cell is shifted to a richer medium, which doubles its growth rate. Then the cell will reach the volume $2V_r$ at 30 minutes, before chromosome replication and septation $(C+D)$ are completed. The cell cannot then divide at volume $2V_r$, since chromosome replication and septation have not finished, and will have reached a considerably larger volume by the time it is permitted to divide. Therefore when the daughter cells do separate, they will be considerably larger than V_r. Their size will in fact be

$$\frac{V_r 2^{(C+D)/30}}{2}$$

instead of V_r. Then the volume at birth of a cell growing with a generation time g, referred to as the reference volume, will be

$$V=\frac{V_r 2^{(C+D)/g}}{2} \qquad (3.1)$$

Evidently, the faster the growth, the larger the cell volume, as discussed in the last chapter in relation to the studies of Maaløe and his colleagues on *Salm. typhimurium.*

Using the model of Pritchard *et al.*, we can take the argument one step further to find a relationship between V_r and the concentration of the initiation repressor molecules in the cell, H. Assume that the number of such molecules produced per chromosome is a. The reference volume is again V_r, as before, which is the volume of a cell which has only one replication pair which initiates with every division. Then the concentration of the repressor produced is

$$h_c = \frac{a}{V_r} \tag{3.2}$$

since the amplitude change is h_c. Substituting for V_r from equation (3.2) into equation (3.1) we get

$$V = \frac{a}{2h_c} 2^{(C+D)/g}. \tag{3.3}$$

Thus we have relationships connecting cell size, generation time, reference volume and threshold concentration of initiation repressor. These are made possible by the fundamental observations of Helmstetter and Cooper on the constancy of $C+D$, which is valid over a considerable range of generation times although it breaks down for very slowly growing cultures.

There are obviously many problems which remain untouched by these considerations, which do no more than lay down some basic rules of behaviour in the prokaryotic cell cycle. The relationship established between generation time and volume begins to bring spatial considerations into the control picture, though in a minimal manner. If we wish to draw a diagram of the control circuit for DNA replication, we must somehow devise a method of introducing the growth process itself, which has been unnecessary up to now. Using the conventions introduced in the last two chapters we could consider a diagram of the sort shown in Figure 3.3 to represent the logic of the Pritchard model according to feed-back principles. Here we show growth as the 'end product' of the action of RNA and protein on nutrients. The circle with R_h is mRNA for the initiation repressor, H, which blocks DNA replication, designated by a double arrow. Growth inactivates H by the simple process of dilution. Clearly this is an unsatisfactory representation because the effect of growth on H is not mediated by a specific signal. The use of control circuit diagrams is of value and can be unambiguously interpreted only if arrows correspond to identifiable signals or messages mediated by specific molecular species. We have now had to break this convention.

In deriving differential equations to represent the dynamics of such a model, the effect of growth can be taken account of in a relatively simple manner. Let us write the concentration of a molecular species such as a

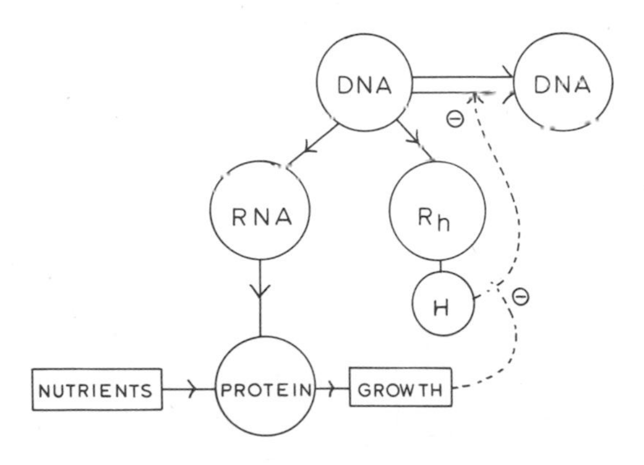

Fig. 3.3. A control scheme for the relationship between growth and DNA replication in bacteria, based on the model of Pritchard *et al.* (1969).

particular type of mRNA in a cell of volume V in the form

$$X = \frac{N_X}{V}$$

where N_X is the number of molecules of the messenger in the cell. Similarly, let us write the concentration of the homologous protein species as

$$Y = \frac{N_Y}{V}$$

The time variation of Y is given by

$$\frac{dY}{dt} = \frac{1}{V}\frac{dN_Y}{dt} - \frac{N_Y}{V^2}\frac{dV}{dt}. \tag{3.4}$$

The number of these protein molecules synthesized per unit time may be taken to be proportional to the number of homologous messenger molecules present, which is $N_X = XV$, assuming as before that messenger is the rate-limiting factor for protein synthesis. Therefore, if protein is stable (no degradation),

$$\frac{dN_Y}{dt} = \alpha XV.$$

Substituting this in equation (3.4) and writing Y in place of N_Y/V, we get

$$\frac{dY}{dt} = \alpha X - \frac{\dot{V}}{V} Y, \qquad \text{where } \dot{V} = \frac{dV}{dt}.$$

If growth in cell volume is exponential, then

$$\dot{V} = \mu V$$

and equation (3.4) reduces to

$$\frac{dY}{dt} = \alpha X - \mu Y. \tag{3.5}$$

For growth which is not exponential, the differential for changes in protein concentration would involve time explicitly. If protein is unstable so that it is broken down at a rate proportional to its own concentration, there will be an additional term, $-\gamma Y$, in equation (3.5), so that it takes the form

$$\frac{dY}{dt} = \alpha X - \beta Y, \qquad \text{where } \beta = \mu + \gamma$$

Similar considerations apply to the other molecular species, so one sees that the general form of the equations is not necessarily altered by the occurrence of growth, which, if exponential, enters as an additional parameter. In the next chapter I will consider equations both for single cells and for cell populations which allow one to study the details of the cell cycle under varying conditions of culture. And in the last chapter an analytical treatment of growth, seen as a co-operative process, will be presented.

Eukaryotes

The cytological basis of the distinction between prokaryotes and eukaryotes is the possession by the latter of a defined nucleus, separated from the cytoplasm by a nuclear membrane. Why such an organelle should have evolved is not obvious. Gene activity and regulation can occur perfectly well in a cytoplasmic milieu, as occurs in prokaryotes, and it could be argued that the existence of a nuclear barrier acts simply as a hindrance to the free passage of information between the genes and the rest of the cell. Certainly from the control point of view the nuclear organelle is not easily understood. However, it is useful to take account of structural considerations, such as the packaging problem for the safe delivery of replicated chromosomes to daughter cells. If chromosomes were scattered about the cell interior at random, it would be more difficult to bring them together for the sorting operation of mitosis. This problem could be solved by the same method as that used in bacteria, attachment of chromosomes to membrane sites and their physical separation by membrane growth, but this operation is much more difficult in a cell without a defined axis. Furthermore, the absence of a wall for structural support in animal cells and the multiplicity of the chromosomes makes such a process more difficult still. So the construction of an internal sorting mechanism, the mitotic spindle, becomes something of a necessity and the nuclear envelope keeps the chromosomes together as a package to ensure that none gets lost before the essential separation operation begins.

However, there may be a more significant role for the nuclear membrane in eukaryotes in relation to the process of gene regulation itself. The evolution of complexity in the genome of higher organisms appears to have occurred by reiteration of existent genes and the subsequent divergence of base sequences in the copies. Not all the RNA transcribed from these copies is necessarily functional and there is a very large amount of RNA which is broken down within the eukaryotic nucleus without ever apparently serving any informational or structural purpose. This continuous turnover of RNA within the nucleus may be a necessary aspect of the adaptive potential of these organisms, and if so the nuclear membrane serves the important role of containing the RNA so that it can be efficiently re-cycled and maintained in relatively high concentration. A similar argument can be applied to the problem of keeping control molecules such as repressors close to their sites of action without getting lost in the considerably larger volume of the eukaryotic cell. The actual nature of the regulatory molecules in eukaryotes is a subject of active debate, but Naora (1973) has presented convincing evidence that low molecular weight nuclear RNAs are strong candidates for a regulatory function. This position has been amplified by Reanney (1975) in a provocative theory about the role of RNA in metazoan regulation and embryonic induction, wherein he gives an explanation of the reason for the rapid metabolism of RNA within the nucleus. Undoubtedly there are many physiological roles which the nuclear membrane plays, including other selective permeability functions, but the current state of understanding of these processes is still fragmentary and undergoing rapid change. Certainly no control picture of any degree of reliability can be constructed for the eukaryotic cell at this stage of our understanding. However, this will not deter us from attempting to extract some simple logical principles, and proceeding with a general analysis which does not depend upon molecular detail.

The Compartment Picture of the Cell Cycle

The mitotic cycle is conveniently described as a temporally ordered sequence of physiological events which together constitute the total set of processes occurring between one mitosis and the next. We saw that in bacteria there was a well-defined ordered relationship between the replication of the chromosome and the growth of the septum which physically separated the two daughter chromosomes, enclosing them in different cells. Such temporal ordering is necessary for the fulfilment of structural requirements: for example, chromosomes must be duplicated before they can be partitioned to daughter cells. It is the existence of structural constraints of this type that makes necessary the occurrence of temporal ordering processes in cells, periodic signals which arise from metabolic oscillations of some kind. In prokaryotes we found that successive rounds of chromosome replication could overlap one another in time, since a second initiation event could occur before the first chromosome replication was finished. No structural constraints are violated by such overlap. However, in eukaryotes the phase of chromosome duplication is always temporally separated so that only one doubling can occur in any cycle (except for the special circumstances which lead to polyploidy, when there are successive chromosome duplications without separation). This allows one to represent the cycle as shown in Figure 3.4. The most easily identified parts of the cycle

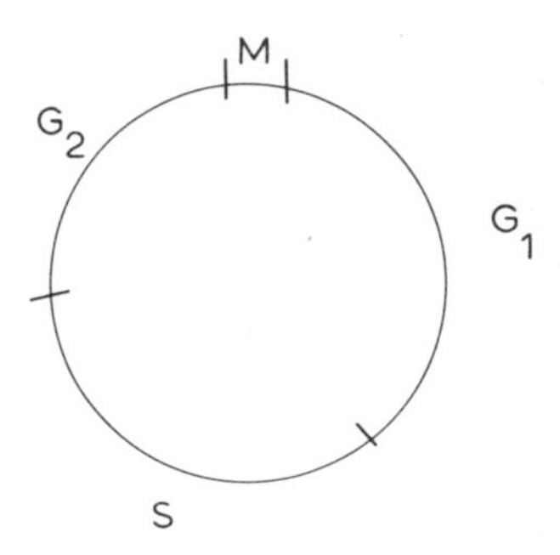

Figure 3.4. The phases of the eukaryotic cell cycle.

physiologically are M, the phase of mitosis; and S, the DNA replication phase. These phases or compartments are separated by others known as G_1 and G_2, during which other events occur about which we still have very little knowledge (hence they are designated as gaps). G_1 is subject to great variation in its duration, even for one particular cell type such as epidermal cells of a mouse, while S and G_2 are usually more strictly defined, each lasting several hours, although there are exceptions to this as well (e.g., in *Hydra* it is G_2 that is highly variable; David and Campbell, 1972). The phase of mitosis or nuclear division is short relative to the others, and it is usually followed by cytokinesis, the division of the cell into two, not shown in the diagram.

The problem facing the experimentalist concerned with an understanding of the eukaryotic cell cycle is to provide some account of the correlations between the events occurring in the different compartments and then, should it suit his fancy and his philosophy, to construct a causal model to account for the passage of a cell from one to the other. The compartments are of course cell states, defined in terms of the concentrations of the various substances, their physical relationships, and the rates of the different processes occurring at any one time. Obviously the extent to which each compartment can be itself subdivided into constituent states or sub-compartments depends upon the detailed information available about what goes on within them. So far this is meagre.

Saccharomyces cerevisiae

Important insight into the organization of events in the eukaryotic cell cycle is provided by an interesting genetic study of yeast by Hartwell *et al.* (1974). They isolated temperature-sensitive mutants which had normal growth at 23 °C but were arrested at some specific point of the cycle at 36 °C. These were classified according to the cell cycle events which they were capable of carrying out and a picture of the causal relationships between the processes observed such as initiation of DNA replication, DNA synthesis, bud emergence, nuclear migration, cytokinesis, etc., then began to emerge. The investigators concluded that there are two causal chains in yeast, one concerned with intranuclear events such as initiation of DNA replication, DNA synthesis and nuclear division, and the other involving bud emergence, nuclear migration, cytokinesis and cell separation. Normally these are coupled together at the beginning by a common 'start' event and at the end by cytokinesis and cell separation. However, there are mutants in which there is an uncoupling so that, for example, successive rounds of intranuclear events occur in the absence of bud emergence and cytokinesis; and, conversely, a mutant in which successive rounds of bud emergence occur at the normal frequency without any correlative nuclear events. This is very similar to the uncoupling of chromosomal and cytoplasmic events which can occur in bacteria, successive rounds of DNA replication being possible in the absence of septation and cell separation, and successive rounds of growth and division occurring in certain temperature-sensitive mutants in the absence of DNA synthesis. One is thus led to the view that either there is a separate oscillator or clock controlling each periodic sequence, the oscillators being normally coupled together; or that both are driven by a separate oscillator, which would then govern the 'start' event.

Physarum polycephalum

A very useful organism to use in the study of the causal relationships defining the mitotic cycle is the true slime mould *Physarum polycephalum*. In its vegetative phase this organism exists as a syncytium, a continuous mass of cytoplasm which can be several centimetres in extent and contains millions of nuclei. Under laboratory conditions, the moulds are usually grown on filter paper impregnated from below with nutrient medium, and are restricted in size to two or three centimetres across their diameter. The feature of this organism which lends itself admirably to biochemical studies on the mitotic cycle is that the continuity of the cytoplasm results in a high degree of mitotic synchrony among all the nuclei, so that one is effectively dealing with a system that behaves like one very large cell. The large size is of course important for biochemical analyses. A comprehensive review of the behaviour of this organism and of the pattern of biochemical events characterizing its mitotic cycle can be found in Rusch (1970). The information I want to extract from this for the purposes of constructing models of the mitotic cycle is restricted to some elementary observations about necessary conditions defining some of the relationships between events occurring in the different compartments of Figure 3.4.

Plasmodial cultures of *Physarum* are initiated by pipetting an aliquot of microplasmodia, moulds grown in submerged culture conditions in reciprocating flasks whose agitation prevents fusion, on to filter paper where the small moulds coalesce and form a large syncytial mass. Within each microplasmodium mitoses are synchronous, but there is no synchrony between them since there is no direct communication. Thus fused microplasmodia are initially asynchronous, but by about ten hours after coalescence, 99% synchrony of mitosis is observed through the mould. The general variability of G_1 in eukaryotes is emphasized by the fact that this phase does not exist at all in *Physarum*; DNA synthesis starts immediately after telophase, the last stage of M, and continues for about 3 hours. This is followed by G_2, which has a duration of 5 to 7 hours at 26 °C when the mould is grown on semi-defined medium. Mitosis lasts for about 20 minutes, so the total cycle time is eight to ten hours.

One can ask whether DNA replication is a necessary condition for mitosis in *Physarum* and get a rather clear answer. Nuclei in small plasmodia in S-phase cannot enter mitosis when fusion occurs with a large plasmodium just ready to enter mitosis, indicating that chromosome replication must be completed before mitosis can occur. This is reinforced by the observation that the delay of DNA replication by inhibition with 5-fluorodeoxyuridine (FUDR) delays mitosis by an equal period. Release of the nuclei from inhibition resulted in mitosis about 6 hours after the release, indicating that a 6-hour G_2 is also a necessary part of the preparation for mitosis. Nuclei that have completed DNA synthesis can enter mitosis immediately however, if transferred to late G_2 plasmodia, so the preparation in G_2 is primarily cytoplasmic. This state of preparedness for mitosis can be inhibited by heat shock (30 minutes at 38 °C), which causes a delay proportional to the time the mould has been in G_2, up to about 20 minutes before M. After that, the heat shock results in an abortive mitosis due to disruption of the mitotic spindles, chromosomes fail to separate and nuclei enter S-phase again to become tetraploid after the pseudo-mitosis.

Cultures initiated at different times and so at different phases of the mitotic cycle can be made to coalesce with one another so it is possible to carry out fusion experiments to look for evidence of dominance relationships between different stages of the cycle. If one entertains an initiator-type hypothesis for control of DNA replication, one might anticipate that fusing a mould in S-phase with one in G_2 might result in the premature initiation of DNA replication in the latter nuclei. However, this did not occur, an observation that is at variance with similar experiments performed with *Amoeba proteus* (Prescott and Goldstein, 1967) and with fusions of HeLa cells and hen erythrocytes (Harris, 1967). If one assumes the initiation repressor type hypothesis, then one would anticipate that if nuclei in S-phase were transferred to a mould in G_2, they might cease DNA replication, since they would now be exposed to a super-threshold concentration of the repressor. This expectation was not realized either; S-phase nuclei continued to incorporate [^{3}H]thymidine in G_2 cytoplasm. However, a closer examination of the hypotheses suggests that these observations are not unexpected. Initiator protein, if it exists in *Physarum*, would be used up or destroyed soon after synthesis, so that S-phase moulds would not necessarily have extra initiator after the very early stages of DNA replication. Similarly, initiation repressors would be effective in inhibiting only those events associated with the very beginning of S-phase. Once chromosomes have started replication, they could well behave autonomously. There is indeed good evidence that there is a highly ordered sequence of events governing the replication of each single chromosome of the eukaryotic nucleus, different parts of the chromosome being replicated in a defined sequence. These chromosomes are of course much more complex than the bacterial chromosome, consisting of many (about 10^3) individual replicating units or 'replicons', each of which is probably synthesized as a unit. The units within one eukaryotic chromosome start their replication in a defined sequence and this may occur autonomously once the signal for the replication of the chromosome has first been received. This first signal may be of brief duration throughout the plasmodium, due either to an initiator substance rising above, or to a repressor dropping below, threshold. Careful timing experiments are required to explore these possibilities. Such experiments have recently been reported by Sachsenmaier, Remy and Plattner-Schobel (1972) and by Kauffman and Wille (1975), providing important insight into the nature of the mitotic control system in this organism.

Sachsenmaier *et al.* used plasmodia whose phase differences were always 6 hours, one-half of the 12 hour cycle time of their cultures and fused them at different phases of the cycle. They discovered that if one plasmodium, say A, was 0 to 6 hours past mitosis and a second, B, was 6 to 12 hours past mitosis, then fusion always resulted in a delay of B and an advance of A. The magnitude of the delay was always about the same, roughly about 3 hours, half the time difference between the cultures. If, on the other hand, plasmodium A is 6 to 12 hours past mitosis and B is 0 to 6 hours past mitosis, the result of fusion is a delay in A and an advance in B, as expected, since the positions of A and B are simply reversed. From these observations, Sachsenmaier *et al.* deduced that mitosis is controlled by a diffusible substance whose concentration varies in a saw-tooth manner, as does the initiation repressor concentration in the model advanced by Pritchard *et al.* (1969) for control of initiation of DNA synthesis in bacteria. However, Sachsenmaier and his colleagues supposed that the substance controlling mitosis in *Physarum* is a mitogen which reaches its peak at mitosis and then falls rapidly to its minimum, being either used up or destroyed during mitosis. Thus their hypothesis is one of positive control of mitosis, analogous to an initiator theory for the control of DNA replication. The reason for this postulate comes from earlier work by Brewer and Rusch (1968) in which it was shown that 30-minute heat shocks (38 °C) applied to cultures of *Physarum* growing at 26 °C caused delays in the time to reach mitosis, increasing linearly from a 15-minute delay at 6 hours before mitosis to a 2-hour delay at 2 hours before mitosis. The delay then dropped off sharply. These observations suggested that a heat-labile substance, presumed to be a protein, accumulates in *Physarum* as the mould prepares for mitosis, and is used up at mitosis; hence the mitogen postulate.

Neither the mitogen hypothesis nor the saw-tooth (relaxation) oscillation hypothesis is a necessary deduction from these observations. The heat shock observations could result from the inactivation of an enzyme which degrades a small molecule acting as a co-repressor in the control of mitosis, for example. In this respect it would be interesting to carry out synchronization studies on moulds which are separated by filter paper with defined pore sizes in order to get some idea of the molecular weight of the communication signal.

The postulate that the wave-form of the oscillator is of saw-tooth or relaxation type arises from the phase-delay to phase-advance discontinuity of moulds on either side of mitosis, as described above, but such a discontinuity is expected to arise from fusion studies of a system with any periodic control signal, providing only that it has the stability characteristics of a limit cycle. This conclusion arises from an elegant qualitative study of the behaviour of limit cycle oscillations carried out by Winfree (1970, 1971) in the context of work on biological clocks, studies which will be discussed in Chapter 4. Kauffman and Wille (1975) have exploited this analysis in their work on *Physarum* and have extended the argument to provide an experimental method of identifying some basic characteristics of the *Physarum* mitotic oscillator. Although the basic principles of this analysis will be described again later in a discussion of Winfree's work on the *Drosophila* eclosion clock, it seems worth presenting the analysis at this point in its application to *Physarum* since the ideas are still relatively unfamiliar ones in biology and I believe that they are of very basic importance in the study of periodic processes.

As defined in Chapter 1, a stable limit cycle is a closed curve which is the asymptotic limit approached by trajectories of a time-dependent process. An example of a limit cycle in two dimensions is shown in Figure 3.5, where one sees trajectories winding out to the closed curve, *C*, from the unstable

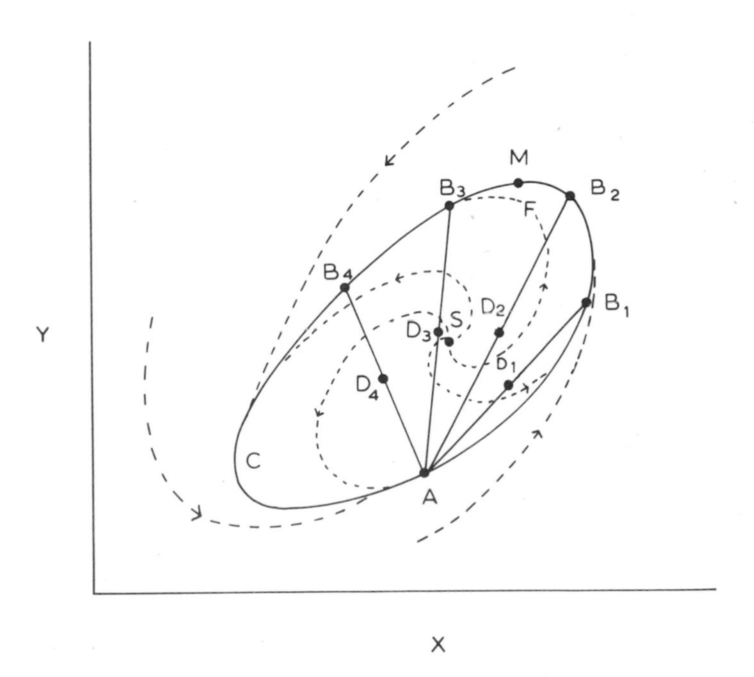

Figure 3.5. The limit cycle with associated trajectories assumed by Kauffman and Wille (1975) to control the mitotic cycle in *Physarum polycephalum*, and the predicted results of fusion experiments.

singularity, S, and trajectories winding in to the cycle from outside. In applying this picture to the mitotic cycle of *Physarum*, we identify X and Y as the two major molecular species involved in the oscillation controlling the initiation of mitosis, one of which could be the postulated mitogen. We have two variables because this is the minimum number which can generate a limit cycle. Even the oscillator of Figure 3.2 is in reality a limit cycle, since there are two variables which are used to describe the process, mRNA and repressor. If one were to plot one of these against the other, the result would be somewhat as in Figure 3.6. The fact that mRNA is assumed to be synthesized in a very rapid burst and then decays means that the parts of the cycle marked 'fast' are traversed quickly while the other part represents the slow dilution of repressor as a result of cell growth.

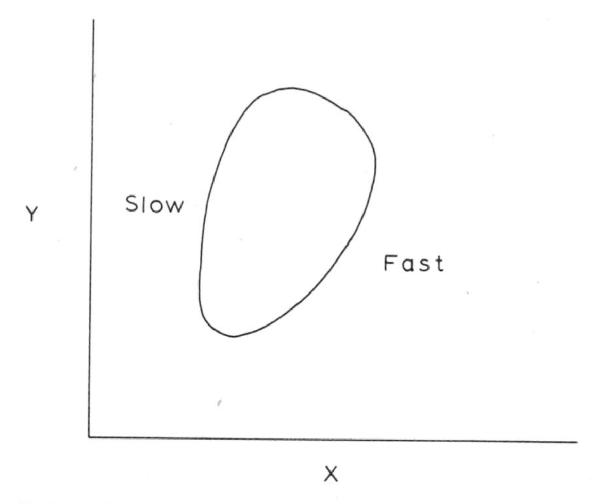

Figure 3.6. The limit cycle controlling the bacterial cell cycle in the model of Pritchard *et al.* (1969), where X = mRNA for initiation repressor (H) and Y = H.

Consider how one would represent fusion experiments between moulds of different relative age in relation to Figure 3.5. Let the point A represent the state of the mitotic control oscillator at a particular time before mitosis. If this is fused with a mould whose representative point is at B_1 (say closer to mitosis, shown occurring at M on the cycle), then their cytoplasmic contents will fuse and they will then find themselves at some intermediate state between A and B_1 represented by D_1, a point on the chord AB_1. This is of course only a qualitative geometric representation. The fused moulds will then follow the trajectory which passes through D_1 and will rejoin the limit cycle at some point intermediate between the points which represent the states of the original moulds A and B_1, had they continued on their cycle undisturbed. Thus if A and B_1 are, say, 3 hours apart, with B_1 ahead of A, one would expect the fused pair to enter mitosis about one and a half hours behind the expected time for B_1, assuming D_1 to be equidistant from A and B_1.

Suppose now A is fixed so that one mould is always at the same state when fusion is carried out, but the other one is varied, represented by the points B_2, B_3 and B_4. Then the same argument can be used to predict roughly what the effect of fusion will be. Soon after fusion the moulds will be at states D_2, D_3 and D_4. The mould at D_2 will return to the limit cycle on the trajectory passing through D_2 and this will carry it to a point between the expected states of moulds A and B_2. This would again represent a phase delay of B_2 and an advance of A. However, the case of A and B_3, giving D_3 on fusion, will result in a return to the limit cycle at a point between A and B_3 on the opposite side of the singularity, S, from D_1 and D_2 relative to the reference point A. In this case one would say that A has been delayed relative to mitosis and B_3 advanced; and similarly with A and B_4. A discontinuity has thus occurred which arises from the existence of the singularity, S, the (unstable) focus of the limit cycle. Furthermore, the exact point at which the discontinuity occurs allows us to deduce something about the shape of the limit cycle and the location of S, since it can tell us how symmetric or asymmetric the cycle is in relation to the focus. The time spent on the different parts of the cycle can also be estimated by this procedure. Kauffman and Wille (1975) have called this the arc discontinuity and by its use they have deduced the shape of the postulated mitotic control oscillation in *Physarum* (see also Kauffman, 1974). They then used this to predict the effect of heat shocks of varying duration at particular phases of the cycle, showing an impressive consistency with their model. They also succeeded in showing that the oscillator is not a part of the mitotic process itself, since disturbance of the latter can occur without effects on the former.

It should be remembered, finally, that the mitogen postulate is not a necessary one, and that if a repressor model had been assumed then the oscillator would be the mirror image of the one drawn by Kauffman and Wille. Isolation of the control molecule or fusion studies across filters are needed to get further information on this point and others which arise from these very interesting investigations.

The molecular nature of the postulated mitogen has been the subject of recent work by Bradbury and his colleagues (Bradbury *et al.*, 1974*a,b*). Their studies of change in the phosphate content of FI histone extracted from *Physarum* nuclei showed a marked oscillation with a peak at the end of G_2 just prior to mitosis. This pattern coincides with the capacity of sonicated nuclear preparations to phosphorylate FI calf thymus histone, thus strongly suggesting that a nuclear kinase is synthesized towards the end of G_2 and is then rapidly degraded during M- and early S-phase. Heat shock reduces the phosphorylating activity markedly, and Bradbury *et al.* (1974*b*) found good agreement between these temperature effects and the earlier heat shock studies of Brewer and Rusch (1968). The postulate advanced is that phosphorylation of FI histone initiates chromosome condensation and the events of mitosis then follow. The only discrepancy between this proposal and the observations of Kauffman and Wille (1975) is that the latter investigators found that they could disturb the mitotic process without phase-shifting the mitotic clock, suggesting that a nuclear kinase cannot itself be a component of the oscillator controlling the cell cycle, although it must be regulated by such an oscillator.

There are many more details known about the relationships between DNA, RNA and protein synthesis in *Physarum* which will be of great value in constructing a detailed picture of the mitotic cycle in this organism. However, for our present purpose it is sufficient to recognize both the intricacy of the interaction and the difficulties associated with the identification of the physiological signals which regulate the transitions from one compartment to the next. It is important to realize that the very complexity of the interactions result in the possibility that there is no fixed set of signals which always control the transitions, but that under different conditions different necessary conditions can become rate-limiting. Thus it may be better to look upon the cell cycle as a mixture of probabilistic and deterministic processes, with some overall constraints about temporal ordering, but also many events which can occur in random sequence. Some of the transitions from one compartment to the next may be governed by probabilistic rather than deterministic events. This means that, instead of considering models of the type shown in Figure 3.2, which is a deterministic oscillator, we might consider purely random processes of, say, Poisson type, which determine the transition rate from one state of the cycle to another. I shall consider such a model shortly, but at the moment it is worth observing that there is no necessary discrepancy between these. The deterministic oscillator of Figure 3.2 needs to be embedded in a 'noisy' biochemical space before it can approximate to the real behaviour of the process governing the initiation of DNA replication which has a variability of 10–20%. I have considered such a model for the bacterial cell cycle (Goodwin, 1970), but there were difficulties with the application of statistical mechanics to the limit-cycle there used to represent the oscillator. It is simpler and more satisfactory to use the original model which I explored in my book and discussed briefly in the last chapter, taking account of the results of Fraser and Tiwari (1974). I will make use of this in the next chapter to explore interactions between eukaryotic cells relating to the control of growth. At this point, I would like to pass through the levels of eukaryotic complexity to the metazoa, and in particular to the mammal, where not only does one have the life cycle to consider but its relation to tissue homeostasis, including the phenomenon of cell differentiation and the problem of neoplastic growth.

The Mammalian Cell Cycle

Many and varied are the cell types which have been employed in studying the ordering of events during the growth and division cycle of mammalian cells, from baby hampster kidney to the HeLa cell of human cervical tumour origin. A great deal of detailed information has accumulated and some general models have emerged which tend to guide contemporary research programmes. One of these involves an interesting speculation about the relationship between DNA replication and gene control which has been called the 'clean gene' theory. In 1964, Kishimototo and Liberman, working with rabbit kidney cells, observed that if DNA-dependent RNA synthesis was blocked during S-phase by the use of the drug actinomycin D, then mitotis was arrested; but if added one hour after the end of S-phase, mitosis occurred normally. They suggested that some mRNA was synthesized during S-phase and up to one hour after its termination which produced protein required for mitosis. Using puromycin to inhibit protein synthesis, they found that the addition of this drug at any time during S (about ten hours duration) and the greater part of G_2 (four hours duration), caused a complete cessation of mitosis, consistent with their proposition. They made the further interesting conjecture that in mammalian cells, mRNA synthesis may be obligatorily coupled to S-phase. The plausibility of this hypothesis depends upon the observation that both DNA polymerase and RNA polymerase need naked or 'clean' DNA to work on and this may only become available during S-phase, when the DNA gets stripped of its protein coat (histone and acidic protein). This stripping or uncovering does not involve the whole of any chromosome simultaneously, but only the parts undergoing replication.

This hypothesis has not survived in its original form, since it is now known that mRNA synthesis can occur during G_1 and G_2 as well as during S-phase, but a somewhat more general formulation known as the 'quantal mitosis theory', first explicitly formulated by Holtzer (1963), currently exists. This theory postulates that in order for there to be a change in the regulatory state of a gene (i.e. turning a gene on or off), a cell must pass through a mitotic cycle during which the gene can be switched by some control signal. Thus a cell becomes competent to change its epigenetic state only if it passes through a cycle, S-phase being of course significantly, but not necessarily uniquely, implicated in this development of competence. Since in general cells which have undergone differentiation have ceased dividing, the stability of the differentiated state would be nicely accounted for by this theory and it is supported by important evidence in addition to that presented by Holtzer (1963) in relation to the process of myogenesis. For example, Stockdale and Topper (1966) carried out some very interesting studies on the induction of casein production in mouse mammary gland tissue. They found that cells could be stimulated to enter S-phase and proceed to cell division by the addition of insulin to organ cultures of the gland tissue, but that casein production was induced only if hydrocortisone and prolactin were added as well as insulin. The latter two hormones if added without insulin failed to induce casein production. If colchicine was added to cultures together with all three hormones, then again there was no casein production; nor was there a cell cycle, due to the colchicine which blocked cells in mitosis. They were also able to show that under conditions where DNA synthesis and cell division are delayed in response to insulin, as occurs in mammary gland tissue from virgin mice, casein production arising from the presence of hydrocortisone and prolactin is also delayed, both by about 48 hours. Thus the correlation observed is a strong one.

More recently, however, Vonderhaar and Topper (1974) have been forced to alter the original hypothesis of an obiligatory coupling between an S and/or an M-phase and the development of cell competence to respond to a specific inducing stimulus in mammary glands by the observation that explants of tissue from adolescent mice can produce milk proteins in response to insulin, hydrocortisone and prolactin without passing through S or M. Only mature virgin mammary gland tissue was found to require passage around the cell cycle before becoming competent. This suggested to the investigators that mammary gland cells in mature virgin mice may be arrested in G_1 at a phase after some critical period coinciding with that of competence, while cells from adolescent mammary glands may be arrested in G_1 prior to this period. This postulate alters Holtzer's original hypothesis very considerably, requiring the abandonment of the idea that a total cycle (quantal mitosis) is required, or that there is any necessary correlation between the events of S-phase and those involved in gene induction. What remains, however, is the interesting notion that there is a critical period in G_1 when a cell becomes competent to change its epigenetic state. It is suggestive to observe that a number of membrane-mediated properties such as the transport rate of Na^+ in *L*5178Y mouse cells (Jung and Rothstein, 1967) and the uptake rate of 2-aminoisobutyric acid in Chinese hamster ovary cells (Sander and Pardee, 1972) doubles at some time in G_1. A very interesting model which synthesizes a diversity of phenomena, suggesting a primary role for cyclic nucleotides and calcium in cell cycling and differentiation, has been proposed by Berridge (1975).

That a cell cycle is not a necessary condition for differentiation has been clearly established in the case of head regeneration in *Hydra* (Wolpert *et al.*,

1971), shown by blocking cell divisions with u.v. irradiation and getting, nevertheless, full regeneration of hypostome and tentacles. Furthermore, in a vertebrate, *Xenopus laevis* (South African clawed toad), Cooke (1973) has shown that normal morphogenesis and histodifferentiation up to stage 26 occur in embryos in which cell cycles are arrested from stage 10 by the use of the drugs colcemid or mitomycin. Changes of presumptive cell fate induced by the implantation of a secondary organizer also occur normally under these conditions. Such observations are consistent with the observations of Vonderhaar and Topper (1974), but not with the quantal mitosis postulate.

Control of Cycle Rate and Cell Differentiation

A slight alteration of Figure 3.4 is now required to take account of cell differentiation in the metazoa. One generally finds this represented as in Figure 3.7, where there is an exit arrow from the cell cycle after M, indicating that cells at this bifurcation point take a decision whether to enter a pathway of differentiation or to continue on around the cycle. People differ in their opinion about where this departure occurs, whether immediately after M, or at some point in G_1. However, since differentiating cells are in general

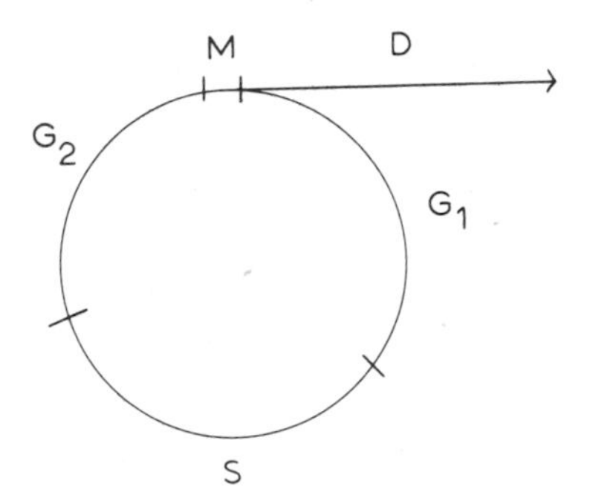

Figure 3.7. The eukaryotic cell cycle, including the divergent pathway to cell differentiation (*D*).

diploid, not tetraploid or with higher ploidy, the normal pathway to differentiation does appear to be after mitosis and not from G_2. This observation of course has its exceptions, as with nearly all biological observations, since organisms use an immense variety of different strategies in attaining their ends.

There is a feature of Figure 3.7 which is distinctly misleading and must be borne in mind as a limitation of the diagram. It is that the process of cell differentiation is not an all-or-none affair in relation to the cell cycle. During embryogenesis the tempo of cell division gradually decreases as organogenesis and cell differentiation proceed, but the two states definitely coexist:

partially differentiated cells (myoblasts, erythroblasts, neuroblasts, etc.) continue to divide. Furthermore, the stem line cells of different tissues for organs in the adult (those which are maintained in a stable state by continuous cell renewal and differentiation, such as skin, liver, kidney, etc.) are distinctively differentiated one from the other, though their role is to undergo division at a rate determined by some criterion of demand by the tissue or organ (to use control language). Despite the fact that in such systems cells do tend to change their state rather abruptly after leaving the stem cell population in response to some signal or stimulus which initiates differentiation, the irreversible all-or-none representation of this process is quite inaccurate. A much more dynamic picture than that provided by Figure 3.7 is required to convey the flexibility of the relationship between the cell cycle and cell differentiation.

I would like now to consider some studies which give one a dynamic picture of tissue organization and at the same time allow the construction and exploration of a negative feed-back model of tissue homeostasis. Having found that the problem of control of DNA replication forced us to consider how cells measure their size, we will now find that the problem of the control of cell differentiation in a tissue forces us to consider how its spatial organization is maintained. This brings us to yet another level of the biological hierarchy, characteristic of the metazoa. We are still operating dynamically within the epigenetic system, although the time scales are now extending into days. This is because the cell cycle time of higher organisms is typically in the range 10–100 hours.

Mammalian Epidermis and the Chalone Theory

A tissue which lends itself admirably to the analysis of the spatial and temporal relationships underlying its stability is the mammalian epidermis. It is simple in organization, readily accessible and has great powers of regeneration. Essentially, the epidermis consists of a basal layer, one cell in thickness, where divisions occur, a layer two to three cells thick where cells are found to be progressively more differentiated the further they are from the basal layer and a top layer of squames or sheets of fully differentiated, keratinized cells. These squames slough off with wear and are replaced by others arising from the differentiation process.

The tissue is easily wounded and the healing process can then be studied in detail. A primary response to wounding is a local increase in the mitotic index of cells in the basal layer. Such a response could be elicited by either a positive stimulus or by the removal of a negative one, as we have found throughout the analysis of control processes in cells. The theory associated with positive control postulates that a wound hormone is released from damaged epidermis and stimulates basal layer cells to divide. Negative control suggests the existence of an inhibitor of the cell cycle which is decreased locally as a result of wounding. A simple test which distinguishes between particular forms of these hypotheses is the following. A wound is made on one side of the mouse ear, removing the epidermis from a patch a few millimetres across. Since the flesh of the ear is so thin, a wound hormone released from the damaged epidermis at the periphery of the wound might be expected to diffuse not only to adjacent basal cells on one side, but through the corium to basal cells on the other, thus stimulating mitotic activity on both sides at the margin of the wound. On the other hand, if removal of the epidermis results in the removal of an inhibitor, then the basal cells of the unwounded surface might be expected to increase their mitotic rate throughout the area of the wound. The latter result was reported by Bullough and Laurence (1960). This observation stimulated the study of extracts of epidermis for the presence of a mitotic inhibitor (Bullough and Laurence, 1964), which has led to the identification and partial characterization of the inhibitor, named a chalone by Bullough (1962), as a glyco-protein of molecular weight about 30 000–40 000 (Hondius, Boldingh and Laurence, 1968). This chalone had the effect of suppressing mitotic activity of epidermis *in vivo* and *in vitro* by up to 60%, depending upon concentration and other factors which enhance its activity, notably adrenaline and hydrocortisone (Bullough and Laurence, 1968*a*). It is tissue- but not species-specific, the former property being required of a diffusible homeostatic control signal.

Several studies have now been carried out which demonstrate the presence of chalone-type inhibitors in tissues other than the epidermis, as reviewed by Bullough (1975). It should be mentioned that the general principle of growth control of tissues by negative feed-back antedates by many years the specific form of the hypothesis embodied in the chalone theory. Weiss and Kavanau (1957) discussed a general form of this theory and experimentalists such as Saetren (1956) had produced evidence for inhibitors extracted from tissues. Such reports are always received with some scepticism because of the ease with which inhibition of normal cell growth can be caused by virtually any impure tissue extract. A convincing demonstration of a specific inhibitor is therefore dependent upon its purification and the characterization of its action, together with evidence of its specificity. This has not yet been achieved for any of the chalones which have been studied and until a complete biochemical identification has been obtained there will continue to be uncertainty about the status of the chalone hypothesis. However, the continuing accumulation of evidence supporting the general proposition of tissue-specific mitotic inhibitors certainly suggests that chalones will ultimately emerge as important elements in the metazoan cell cycle control complex.

An important question arises in relation to the point of action of chalone in the cell cycle. The original observations of Bullough and Laurence (1964) indicated a control point in late G_2, since release of cells from inhibition resulted in the rapid appearance of an elevated mitotic index in strips of epidermis. In addition, Iversen (1970) found that mouse epidermal extracts inhibited [^{3}H]thymidine incorporation into HeLa cells, suggesting that there may also be a control point in late G_1. More recently, Marks (1973*a*) and Elgjo (1973) were able to distinguish two inhibitors in mouse epidermis, one of which depressed the mitotic index and the other the tritiated thymidine index of epidermal cell cultures. The latter, called a G_1 inhibitor, appears to be associated with the cell membrane. Both molecular species are glycoproteins, but they differ in their biochemical characteristics. Elgjo (1973) found that the G_1 inhibitor is present mainly in differentiating cells, while the G_2 inhibitor is primarily in the basal cells. A study of Frankfurt (1971) suggested that, apart from its inhibitory effects, epidermal chalone induces cells to differentiate and can even cause the passage of cells from G_2 to the differentiation pathway without passing through M.

The very large number of possible molecular control circuits whereby chalone could exert its effects on gene activity in the epidermal cell nucleus makes it unprofitable to go into this question in any detail. A number of different theories of gene control in eukaryotes have been proposed, such as those of Britten and Davidson (1969), Crick (1971), Paul (1972) and Naora (1973), but their details are not really of importance for our present problem which is dynamic rather than molecular. The objective is to demonstrate how a particular realization of the chalone hypothesis provides some insight into the homeostatic properties of the epidermis both in space and in time. In order to keep the picture simple initially, let us use prokaryotic conventions in describing the organization of the genetic control circuits in epidermal cells, while remembering that no implications are being made about detailed molecular mechanisms. In Figure 3.8 the circles labelled cell cycle 'operon' and differentiation 'operon' refer to constellations of genes which are almost certainly scattered over several chromosomes. Their activity is required for active cell cycling or a particular differentiated condition, respectively.

Let us assume that the specific condition of cells for a particular tissue includes two properties. (1) All the differentiation operons other than the specific one for the differentiation of that tissue, the *i*th, are permanently turned off. (2) The cell is in a state such that only chalones specific to that tissue are produced and only G_1 chalone of type i is capable of entering the membrane and activating the metabolic process shown producing the metabolite M_1 which represses the genes of the cell cycle operon. Chalone could, of course, act directly upon the cell cycle operon, but it seems more in keeping with current ideas about metabolic modulation to have some intracellular messenger, M_1, carry out this repression function.

G_1 chalone production is assumed to result from the activity of a specific chalone gene in the differentiation operon. The cell cycle operon includes a

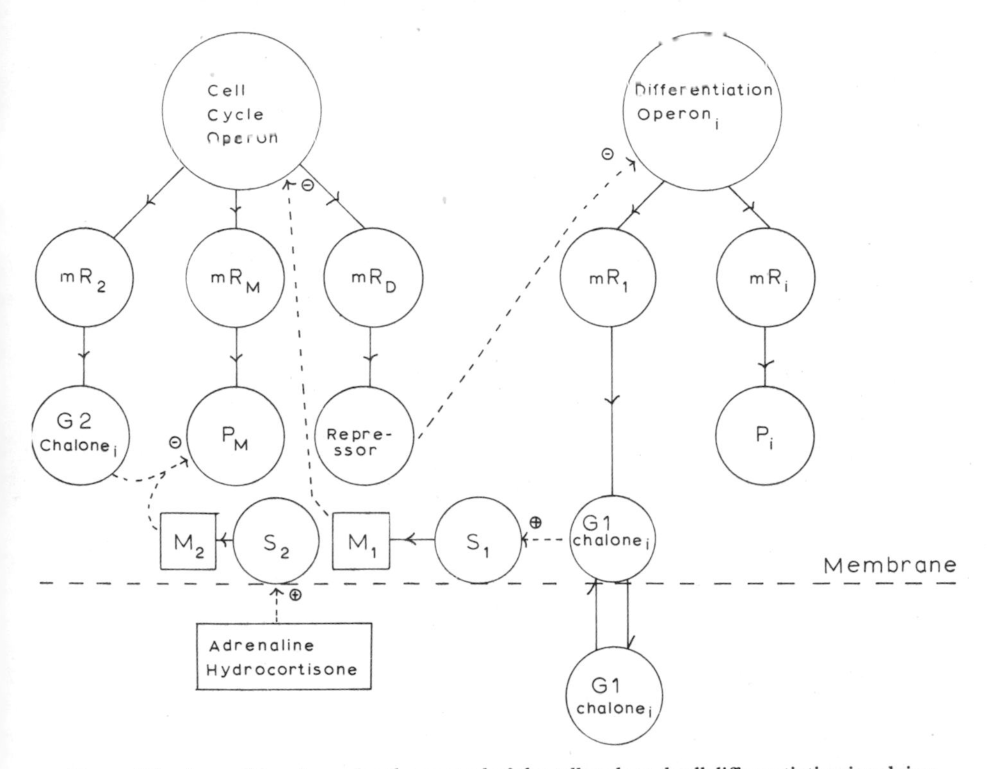

Figure 3.8. A possible scheme for the control of the cell cycle and cell differentiation involving chalones.

gene coding for a 'repressor' which acts specifically on the differentiation operon and one for the G_2 chalone specific to that tissue. This chalone is shown acting on the proteins required for mitosis, P_M, producing the G_2 block. Adrenaline and hydrocortisone act synergistically with the G_2 chalone via the activation of a membrane-bound sensor S_2 which produces an intracellular messenger, M_2. There is some evidence that this may be cAMP (Marks, 1973*b*). The protein products of the differentiation operon are shown in Figure 3.8 as P_i, produced by their respective messenger RNAs, mR_i. In the epidermis, keratin will be one of these proteins.

The assumptions made about the state of proliferating cells in a tissue such as the epidermis are very strong ones, and they mean that we are totally ignoring the problem how they got into that specific condition to begin with; i.e. we are looking only at the very terminal aspects of development. In Chapters 5 and 6, I shall be considering questions concerning the nature and origin of early developmental fields and a rather different picture will emerge for the processes initiating spatial order from those being considered now, which relate to the maintenance of established tissues. The limited regenerative potential of mature mammalian tissues suggests that early developmental processes differ in important respects from later ones, as will be discussed in Chapter 5. Our concern now is with the problem of tissue homeostasis. Let me repeat that the highly simplified picture of Figure 3.8 is not to be taken literally. The point is to consider the dynamic behaviour of a system with this general pattern of interactions between groups of genes concerned with different cellular functions, and with the environment via specific proteins which can respond to extracellular signals and alter intracellular states.

If we start with the cell in cycle, then the repressor prevents cells from differentiating and cycling will continue. However, the entry of some chalone into the cell membrane will reduce the tempo of the cell cycle, the concentration of repressor will decrease and some intracellular chalone of type i will be produced and incorporated into the membrane. This can give rise to a new steady state which involves continued cell cycling at a reduced rate and partial differentiation of cells. Or the system can undergo progressive change in the direction of differentiation, more and more chalone being produced as the result of the increasing repression of the cell cycle operon and the reduction in the level of repressor, releasing the differentiation operon. The circuitry is such that, depending upon environmental influences and the relative affinities of the control signals, the system can behave either as a switch, flipping into one or other of the states where one operon is on and the other off; or its can stay in any intermediate steady state; or it can oscillate continuously from a condition of growth to a condition of differentiation. Thus the binary logic of Figure 3.7 implying that cells must take a decision at some stage whether to re-enter the cycle or to differentiate, is a considerable over-simplication of the behaviour anticipated on the basis of an extremely elementary circuit diagram. This diagram is itself a simplification and so we should be on our guard about analyses of dynamic processes in terms of binary logic applied to cell states.

Since a primary factor determining the state of a cell is its environment, it is necessary now to take this into account. It is at this point that we must consider the spatial organization of the tissue. A simplified drawing of the epidermis is seen in Figure 3.9, showing the basal layer of stem cells in contact with the basal membrane, cells undergoing differentiation above, and the fully differentiated cornified cells without nuclei and consisting largely of keratin at the surface. The vertically-organized column of cells shown is what Potten (1974) has called an epidermal proliferative unit, hexagonal in cross-section. Cells are in communication with one another within and between these units via desmosomes. Mitotic figures are seen

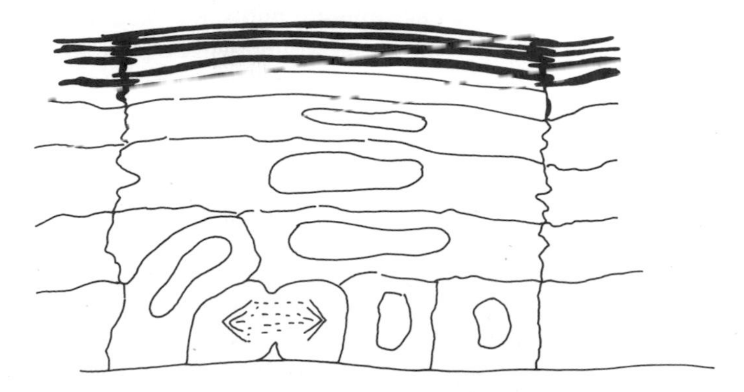

Figure 3.9. A schematic drawing of epidermal structure.

only in cells adjacent to the basal membrane, except after wounding when occasional cells in mitosis are seen higher up. This continuous cell production process at the base, which has a fixed area in an adult organism, results in cells being squeezed out of this layer distally towards the surface, where the squames are continually being lost. So cells travel upwards as they differentiate. If, in accordance with Figure 3.8, the more differentiated a cell is the more G_1 chalone it produces, then there will be a tendency for a gradient of chalone concentration to be established extracellularly, increasing from base to surface. The slope of this gradient will be increased if chalone is also removed by the circulation of blood in vessels beneath the basal membrane.

The other inhibitor, G_2 chalone, produced mainly by basal cells and being quite unstable (Elgjo, 1973), appears to play a role supplementary to the G_1 control process. Its effect may be to produce a small pool of cells in late G_2 which can respond to any disturbance, such as wounding or carginogen application, which increases the rate of release of the inhibitor from basal cells and allows them to divide rapidly, thus assisting the recovery process. It thus plays a local negative feed-back role. Other elements in the overall control system, such as a mesenchymal factor produced by the dermis which induces cell division (Bullough, 1975*a*), have not been considered.

This very simple scheme of Figures 3.8 and 3.9 can be used to explain the essential homeostatic behaviour of the epidermis in response to wounding or other damage. For if a layer of epidermis is removed, either down to the basal membrane or to a more superficial layer, then cells which were producing chalone will have been removed and the local concentration will fall. There will also be an escape of chalone from cells in the vicinity of the wound, due to the absence of other cells, increased circulation, bathing action of lymph, etc. Thus cells will be released from inhibition and the rate of cell cycling will rise. As cells accumulate and are forced away from the basal layer, those at the periphery of the wound will be pushed into lateral contact with differentiating cells producing and releasing chalone, which will induce them to differentiate. These will in turn induce their neighbours, so differentiation should proceed from the periphery to the centre of the wound. But also cells which are pushed away from the basal layer will be displaced further from the sink caused by blood circulation beneath the basal layer, so their chalone concentration will rise and differentiation will proceed slowly throughout the central area of the wound. One might anticipate an overshoot of cell production in the central regions of the wound, where induction to differentiation is slowest. Iversen (1961) has reported overshooting in the mouse epidermis in response to carcinogens from which he deduced the operation of a negative feed-back mechanism of chalone type, but he did not carry out a detailed spatial study. In general, wounds tend to heal progressively from the periphery to the centre.

This simple gradient hypothesis of the spatial organization of a tissue and its stability in response to disturbance is characteristic of a whole class of developmental models for the explanation of pattern formation and morphogenesis. Effectively these are based upon the postulated existence of sources and sinks of a diffusible substance which becomes distributed spatially and determines the state of a cell in any region of the gradient, exactly as is being proposed for the epidermis. In the chalone model each cell is a source and may be a sink if the chalone is broken down at some rate within cells, while the basal membrane defines a boundary with a sink (circulation) on one side and dividing cells on the other. This together with the simple circuitry of Figure 3.8 generates the desired model. Whether or not it has any significant connection with epidermal homeostasis is another question. The epidermis is of course a much more complex tissue than is described in Figure 3.9, which shows only one cell type and its sequence of state changes from cell division to differentiation. In addition, there are hairs, sebaceous glands, pores and many other structures making up the entire tissue. But it is possible that they all operate on principles essentially similar to those described above. An interesting analysis of size and growth control of organs based upon the chalone hypothesis is given by Glass (1973), who demonstrates an instability in such systems which can be interpreted in terms of malignant growth, a subject to be considered briefly below.

The chalone model as described is concerned more with tissue homeostasis than with the stability of single cells, such as the maintenance of a mean cell size in a dividing population. We have not considered this problem of size regulation in eukaryotes as we did in prokaryotes, although it is of course an important one. An early hypothesis, introduced by Hertwig (1908), was that it is the nuclear to cytoplasmic volume ratio that determines when a cell divides. Such a theory could be given a molecular interpretation in terms of either a repressor or an initiator type of control process, with cell volume acting as a dilution factor in the regulation of concentration. For example, the nucleus could act as a source of an initiation repressor and the cytoplasm as a sink (site of its degradation) with rates assumed to be correlated with volumes. Since the nuclear volume remains relatively constant while growth of cell size is primarily cytoplasmic, as growth proceeds the rate of degradation of the repressor will exceed its rate of production and the concentration will decrease. When this has fallen to a critical value, some phase of the cell cycle is initiated (say S-phase) and the cell then progresses to mitosis and division. Such a model, implausible as it is because of its extreme simplicity, would nevertheless provide an explanation for the maintenance of a constant mean cell size for a given growth rate by the same principles as those operating in the Pritchard *et al.* model of bacterial size regulation. And depending upon the assumptions made regarding the relationship between the growth rate and rate of production of the controlling molecule by the nucleus one could have constancy of cell size for different growth rates or some other defined relation between them.

It is now known that in its original form as proposed by Hertwig the model does not conform to the behaviour of eukaryotic cells. Prescott (1956) has shown, for example, that a small nuclear/cytoplasmic volume ratio is not a sufficient condition for the induction of division in the giant amoeba *Chaos chaos*. He showed this by inducing asymmetric divisions in the amoebae and he then observed that the daughter cell with the large cytoplasm and hence a small nuclear/cytoplasmic volume ratio, smaller than that occurring in cells entering mitosis and cytokinesis, did not immediately undergo a second division. However, the time to the next division for such a cell was considerably shorter than the normal generation time, being reduced from 24 hours to 16. These results are entirely consistent with the general picture that emerged from the studies with *Physarum* and rabbit kidney cells, viz. that there are some necessary conditions which must be fulfilled before a cell can divide, one of these being the replication of DNA. They are also consistent with a simple nuclear source/cytoplasmic sink model of cell cycle control, since one would anticipate premature initiation of, say, S-phase if a nucleus finds itself in a large cytoplasm. A chalone-type molecule could play the role of the initiation repressor.

It is often claimed that cell fusion experiments in which one type of cytoplasm induces initiation of S-phase in another type of nucleus such as those described by Harris *et al.* (1969) constitute strong evidence for the existence of a positive signal of initiator type controlling DNA replication. However, it is clear from the above discussion that if different cytoplasms have different capacities to degrade a repressor, then induction by mixing can be explained by either hypothesis. In order to resolve this ambiguity, one needs either control mutants which can be fused with normal cells to study the dominance relationships as carried out by Jacob and Monod on the *lac* operon of *E. coli*; or one must isolate and purify substances which have observable effects on particular events in the cell cycle, as in current research on chalones.

An account of the chalone hypothesis should include some mention of its relation to cancer therapy, for clearly a tissue-specific mitotic inhibitor which can be isolated and purified offers some promise of clinical use in relation to the control of growth in neoplasms. However, a glance at Figure 3.8 presents one with a problem: injection of chalone into an animal with a skin tumour, let us say, should result in the inhibition of mitotic activity in the normal and the neoplastic tissue, but presumably more in the former than the latter, since a tumour has in some sense broken away from the normal physiological control process and would be expected to be less sensitive. As long as chalone injections continue, this suppressed mitotic rate should result, but when injections cease, the *status quo ante* should be restored. This point is argued clearly by Iverson (1970). However, it is very remarkable that a number of observations have now been reported (Bullough and Laurence, 1968; Rytomäa and Kiviniemi, 1968; Mohr *et al.*, 1968) that regression of some tumours can be obtained and a complete cure reported for the case of melanomata in mice and hampsters. These results are of very considerable interest, and they certainly suggest that chalone action must be considerably more complex than is described in the simple diagram of Figure 3.8. A discussion of certain interesting possibilities is given by Bullough (1975*b*) in a recent review article. It is sufficient to point out that the basic concept of control by negative feed-back has evidently been of great and continuing utility not only in relation to prokaryotic organization, but also in relation to one of the perennial problems of metazoan physiology, that of control of cell growth and the complementary escape from control which characterizes neoplasms.

Probabilistic Models of the Cell Cycle and its Control

It was mentioned earlier that whatever model one might choose to construct for the behaviour of single cells during the cell cycle, there must be a stochastic or probabilistic factor entering in some manner even if it is no more than a proposition that the molecular oscillators responsible for rhythmic behaviour in the whole cell have variable phases and frequencies, so that the net result of their activity is a 'noisy' cycle (Engleberg, 1968). Such a model would be deterministic at the molecular level, involving no random effects such as the spontaneous dissociation of repressor–operator complexes or other molecular associations involved in regulation, or fluctuations in rates of reactions arising from thermal effects, collision frequencies, orientation of reacting molecules, etc. Since the molecular picture of biochemical events occurring within the cell is an intrinsically 'noisy' one, especially if relatively few molecules are involved as must be the case for gene control processes, it does seem most plausible to assume that stochastic influences arise primarily from this source. An extreme view would be to suggest that there is no deterministic element of cell behaviour whatsoever in the sense of an oscillator obeying some dynamical constraints which determine a defined wave-form with a particular frequency and amplitude. One could argue, for example, that a cell will divide when it has completed a number of events, each of which has a particular probability of occurrence as a function of time, and that these events can occur in any order. Such a model proposes no causal cycle within cells, since there are no constraints on the ordering of the events. It was considered by Rahn (1932) and leads to a probability distribution for cell generation times which does not fit the data very well, but could be made to by the addition of further postulates such as ascribing different probabilities to the different events. A more serious criticism is that in such a model there would be no defined S-phase followed by G_2 in violation of an overwhelming amount of evidence. A modification of this model was proposed by Kendall (1948), who supposed that the events constituting the cycle, whilst still being governed by time-dependent probabilities, must occur in sequence. One then has a cycle, but there need be no correlations between one cycle and the next, since the process starts anew after every termination and there is no influence of one cycle on the following one. This is not generally the case for a dynamical oscillator, whose behaviour depends upon its initial conditions and the stability characteristics of the process. Since it is known that mother–daughter generation times are correlated as shown by Schaechter *et al.* (1962), by Powell and Errington (1963) and by Kubitschek (1966), some constraint operating from generation to generation needs to be introduced. This can be done in a variety of ways, as described by Marr, Painter and Nilson (1969) in a paper wherein a very general probabilistic model is developed and applied with notable success to data on the bacterial cell cycle.

So far all the descriptions presented of the cell cycle, except that of Rahn (1932), implicitly or explicitly assume the existence of a clock of some kind, a postulate which derives from the sequential nature of the process occurring between one cell division and the next. However, if a crucial step in this sequence is governed by a purely random event, one could legitimately question the basic proposition that growing and dividing cells undergo a cycle. Evidence that this is indeed the case in eukaryotic cells is presented in an interesting and provocative paper by Smith and Martin (1973). They argue that the data obtained for the distribution of generation times in a variety of cultured cell types is strongly suggestive of a random event which controls the passage of cells out of some phase of G_1 and into the deterministic sequence of events leading to the completion of mitosis. This sequence is itself noisy, but is much more highly ordered in time than the random initiating event determining passage out of what they call the indeterminate A state, situated within G_1. Such a proposition modifies in interesting ways the basic concept of the cell cycle by locating most of the noise in one event, rather than having it fairly evenly distributed over the whole cycle. However, we are still left with the problem of constructing a model which explains both the stability characteristics of the growth and division process, as well as its stochastic features. I shall return to this question again briefly in the last chapter in the context of a phenomenological approach to the description of the cell cycle.

Chapter 4

CELL GROWTH, CELL SYNCHRONY AND BIOLOGICAL CLOCKS

It is possible and, in a limited sense, useful to classify three different states of order of cell populations in analogy with the three states of matter: gas, liquid and solid. In the first case, we may consider the properties of a culture of cells each of which has internal order, as we have been considering in the last three chapters, but with no order in the population apart from that which results from an averaging of their individual behaviour. This is the situation that usually prevails in bacterial populations, each cell being a highly ordered system, like a molecule of a gas, but with no interaction between cells strong enough to result in any defined population organization. I will describe a model of stable growth of such a population in the form of the operation of the chemostat and consider the analogy between it and the behaviour of certain types of cell population in the metazoa.

Given the existence of certain dynamical modes within the individuals of a population, it is possible for order to emerge globally in either of two ways: an external influence can operate to cause the individuals to become coherent in their behaviour; or the interactions between the individuals can become sufficiently strong to result in the occurrence of stable dynamical order. I will consider examples of both types of behaviour in relation to the synchronization of cells by means of periodic environmental forcing functions and the persistence of stable rhythms in populations of cells possessing biological clocks. The order observed in such populations is exclusively temporal, individual cells being free to move relative to one another in any way, the only interaction being via some communication signal. This state of partial order of cells may be regarded as analogous to the liquid state of matter, wherein individual molecules are free to move relative to one another but there are certain constraints on the energy distributions.

The third state of organization of cells is where there are spatial as well as temporal constraints resulting in space–time order due to even stronger interactions between the units, in analogy with the condition of molecules or atoms in solids, particularly in crystals. Four-dimensional order characterizes the condition of the developing embryo, which we will explore in the next chapter. One of the things we will discover is that at this level of organization it makes some sense to do away with the notion of cells altogether for certain purposes, since their interactions are such that they can be regarded as a continuum rather than an aggregation of individuals. But in this chapter the single cell continues to play a basic role as an element in the population, whether asynchronous or synchronous.

The Chemostat

A device designed for the growth of cell populations under continuous, chemically-defined conditions was described in 1950 by Novick and Szilard (1950) and by Monod (1950), both advancing essentially the same theory to explain its operation. The apparatus is basically very simple and has been employed since its introduction to great advantage in the study of cell physiology and population behaviour. A diagram of the system is shown in Figure 4.1.

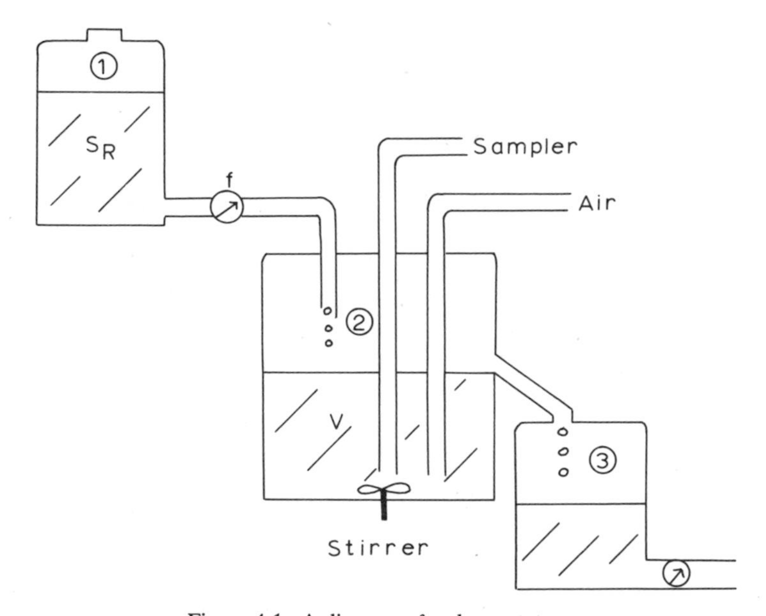

Figure 4.1. A diagram of a chemostat.

Vessel 1 contains the nutrient medium required for the growth of some cell types which can grow in suspension culture, such as bacteria, yeast, LS cells, etc. A control valve shown as a circle with an arrow (usually in the form of a pump running at a certain rate) regulates the rate of flow, f, of nutrient into the growth vessel, 2. This vessel contains a volume, V, of the culture of cells growing on the medium, rapid mixing of new medium being achieved by a stirrer which keeps the contents of the vessel homogeneous. The volume is maintained at a constant value by means of an overflow arm so that the culture flows out into vessel 3 at the same rate as medium enters from vessel 1. An air line and a sampling line into the growth chamber are also provided. Vessel 3 is for collection of cells, which can be used as a source of biological products (drugs, vitamins, protein, alcohol, etc.).

It is intuitively evident that, providing the flow rate through the growth vessel 2 does not exceed the maximum growth rate of the cells within it, this type of system will reach a steady state of growth, which will be stable to small perturbations of cell density or nutrient concentration. To analyse this in detail, let us derive differential equations which describe the behaviour of the relevant variables. These variables are x, the concentration of cells in the vessel, measured in units of, say, mg dry wt/ml of culture; and s, the concentration of the nutrient or substrate which is limiting the growth rate of the cells, also conveniently expressed as mg/ml. It will be assumed that there is such a rate-limiting substrate, for otherwise there is no way of defining what controls growth. The assumption is, then, that some essential nutrient in the medium is maintained at a concentration such that growth rate depends upon it. The concentration of this nutrient in the fresh medium of the reservoir is s_R, a constant.

Consider the equation describing the rate of change of x, cell concentration. Cells will increase by growth, which we will take to be exponential in accordance with the general behaviour of cells growing under constant, defined environmental conditions. Thus the rate of cell growth is

$$\left(\frac{\mathrm{d}x}{\mathrm{d}t}\right)_{\text{growth}} = \mu x.$$

The rate at which cells are washed out of the culture is found as follows. Suppose the flow rate, f, is given in the units ml/min. Then the mass of cells which is washed out of the culture every minute is fx mg/min. This will produce a concentration change in the vessel of fx/V mg/ml. min where V is measured in ml. Therefore this is the dilution effect of washout and we can write

$$\frac{\mathrm{d}x}{\mathrm{d}t} = \left(\frac{\mathrm{d}x}{\mathrm{d}t}\right)_{\text{growth}} - \left(\frac{\mathrm{d}x}{\mathrm{d}t}\right)_{\text{washout}} = \mu x - \frac{f}{V}x.$$

It is useful to define $D \equiv f/V$ so that the equation for x becomes

$$\frac{\mathrm{d}x}{\mathrm{d}t} = \mu x - Dx. \tag{4.1}$$

Next we need an equation for s. This clearly enters the vessel at the rate fs_R mg/min, which produces a concentration change in the vessel of $fs_R/V = Ds_R$ mg/ml. min, where s_R is measured in mg/ml. The wash out rate will be

Ds, s being the concentration of the rate-limiting substrate in the vessel. But there is a third term required in the expression for ds/dt, because cells are growing on s and so are using it up. Define a quantity Y which is the weight of cells formed from one unit of substrate. This is known as the yield constant. For example, for the bacterium *Aerobacter cloacae*, growing on a medium in which glycerol is the rate-limiting substrate, it is found that Y has the value 0·5, so that 1 mg of glycerol is consumed for every 0.5 mg of bacteria produced. Of course the bacteria are taking up many other compounds from the medium as well, such as N, P, S, K, Na, Mg etc., in the form of various salts, but these are present in non-growth-limiting amounts and so they do not enter the equations of the system as variables. The equation for s thus takes the form

$$\frac{ds}{dt} = Ds_R - Ds - \frac{\mu x}{Y}. \tag{4.2}$$

Since we have assumed that s is the variable that limits the growth rate, it is necessary now to define μ as a function of s. We know from the discussion of the operation of the *lac* operon that the uptake of a carbon source into a bacterium is governed by a particular enzyme, a permease located in the cell membrane. Therefore growth rate will depend upon an enzyme-catalysed reaction. This statement would of course be true even if there were no permease system operating, since the rate of utilization of nutrient would again be dependent upon the activity of an enzyme. Therefore a reasonable postulate is that the functional dependence of μ on s should take the form of the velocity of an enzyme-catalysed reaction with s as substrate. At this point we have a choice whether to use the Michaelis-Menten form given by equation (1.9) of Chapter 1, with $K_2 = 0$, or one of the more complex forms (1.11) or (1.23), Chapter 1, describing enzymes with sigmoid kinetics. Novick and Szilard (1950) and Monod (1950) naturally used the simpler Michaelis-Menten expression since co-operative behaviour had not been discovered at that time. So let us follow them and take

$$\mu = \frac{\mu_m s}{K_s + s} \tag{4.3}$$

where μ_m is the maximum growth rate, analogue of maximum enzyme activity and K_s is the analogue of the Michaelis constant $[1/K_1 = (k_{-1} + k_2)/k_1$ in equation (1.9)]. The quantity μ_m is determined by the intrinsic growth rate of the cells and is hence dependent upon their hereditary composition.

Substitution of equation (4.3) into equations (4.1) and (4.2) gives us the pair of equations

$$\frac{dx}{dt}=\left\{\frac{\mu_m s}{K_s+s}-D\right\}x \tag{4.4}$$

$$\frac{ds}{dt}=D(s_R-s)-\frac{\mu_m}{Y}\frac{xs}{(K_s+s)}. \tag{4.5}$$

If a steady state exists, then it will be defined by some pair of values of s and x which balance one another in these equations, both being constant in time so that the derivatives are 0. Designating these values as $\tilde{x}$ and $\tilde{s}$, they must be defined by the equations

$$\left\{\frac{\mu_m\tilde{s}}{K_s+\tilde{s}}-D\right\}\tilde{x}=0 \tag{4.6}$$

$$D(s_R-\tilde{s})-\frac{\mu_m}{Y}\frac{\tilde{x}\tilde{s}}{(K_s+\tilde{s})}=0. \tag{4.7}$$

An obvious solution of equation (4.6) is $\tilde{x}=0$, in which case we find from equation (4.7) that $\tilde{s}=s_R$. This corresponds to the condition of the chemostat when no cells have been introduced into the growth vessel so that substrate is at the same value as it is in the reservoir, s_R. The other solution of these equations is

$$\begin{aligned}\tilde{s}&=K_s\left(\frac{D}{\mu_m-D}\right)\\ \tilde{x}&=Y\left[s_R-K_s\left(\frac{D}{\mu_m-D}\right)\right].\end{aligned} \tag{4.8}$$

It is evident from the equation for $\tilde{x}$ that for a particular value of D, call it D_c, the expression in brackets will be 0, i.e.,

$$s_R-K_s\left(\frac{D_c}{\mu_m-D_c}\right)=0,$$

or

$$D_c=\mu_m\left(\frac{s_R}{K_s+s_R}\right). \tag{4.9}$$

In this case equation (4.4) becomes:

$$\frac{dx}{dt}=\frac{\mu_m s}{K_s+s}-\frac{\mu_m s_R}{K_s+s_R}. \tag{4.10}$$

So long as there are cells in the chemostat, s is less than s_R because the cells must use some of the substrate, and so we see from equation (4.10) that when D has the value D_c, $\mathrm{d}x/\mathrm{d}t < 0$ until $\mathrm{d}x/\mathrm{d}t = 0$, when $s = s_R$. This must then correspond to the solution $\tilde{x} = 0$, $\tilde{s} = s_R$. Thus D_c is the critical dilution rate at which cells are unable to grow sufficiently rapidly to maintain themselves against washout, and so the culture vessel slowly but inexorably empties itself of organisms. In an interesting and informative study of the theory and the experimental behaviour of cells growing in a chemostat, Herbert, Ellsworth and Telling (1956) analysed this situation and showed conditions under which washout occurs. They also discussed many other aspects of the relationship between theory and experiment, both the consistencies, which are many, and the discrepancies, which are of value in suggesting which assumptions of the model may be incorrect. The discrepancies arise mainly in relation to behaviour in the neighbourhood of the critical point, D_c. Cells are better at surviving at high dilution rates than the theory predicts, so the assumption of homogeneity, implying instantaneous mixing of new nutrient with the culture, may be incorrect. Heterogeneities in the environment could account for better survival than expected at the critical dilution rate. Of course any value of D greater than D_c will also eliminate cells.

We can now investigate the stability of equations (4.4) and (4.5) at the two possible steady states. The same procedure as that used in studying the stability of the steady state of equation (1.25) is used. First new variables are defined, $x = \tilde{x} + \xi$, $s = \tilde{s} + \eta$, and it is assumed that ξ and η are small quantities so that all terms higher than linear are ignored. The equations then become

$$\frac{\mathrm{d}\xi}{\mathrm{d}t} = \left\{\frac{\mu_m(\tilde{s}+\eta)}{K_s+(\tilde{s}+\eta)} - D\right\}(\tilde{x}+\xi)$$

$$\simeq \left\{\frac{\mu_m\tilde{s}}{K_s+\tilde{s}} + \frac{\mu_m K_s \eta}{(K_s+\tilde{s})^2} - D\right\}(\tilde{x}+\xi)$$

$$\simeq \frac{\mu_m K_s \tilde{x}}{(K_s+\tilde{s})^2}\eta = a\eta \qquad \text{where } a = \frac{\mu_m K_s \tilde{x}}{(K_s+\tilde{s})^2} \tag{4.11}$$

$$\frac{\mathrm{d}\eta}{\mathrm{d}t} = D(s_R - \tilde{s} - \eta) - \frac{\mu_m}{Y}\frac{(\tilde{x}+\xi)(\tilde{s}+\eta)}{K_s+\tilde{s}+\eta}$$

$$\simeq D(s_R - \tilde{s} - \eta) - \frac{\mu_m}{Y}\left\{\frac{\tilde{x}\tilde{s}}{K_s+\tilde{s}} + \frac{\tilde{s}}{K_s+\tilde{s}}\xi + \frac{K_s\tilde{x}}{(K_s+\tilde{s})^2}\eta\right\}$$

$$= -D\eta - \frac{\mu_m}{Y(K_s+\tilde{s})}\left\{\tilde{s}\xi + \frac{K_s\tilde{x}}{(K_s+\tilde{s})}\eta\right\}$$

$$= -b\xi - c\eta \tag{4.12}$$

where

$$b = \frac{\mu_m \tilde{s}}{Y(K_s + \tilde{s})} \quad \text{and} \quad c = D + \frac{\mu_m K_s \tilde{x}}{Y(K_s + \tilde{s})^2}.$$

The characteristic equation of the system in equations (4.11) and (4.12) is

$$\begin{vmatrix} \lambda & -a \\ b & \lambda + c \end{vmatrix} = \lambda(\lambda + c) + ab = 0$$

so the characteristic roots are

$$\lambda = \frac{-c \pm \{c^2 - 4ab\}^{1/2}}{2}.$$

Since a, b and c are all positive, the real part of the roots must be negative and so the steady state for $\tilde{x}, \tilde{s} > 0$ is stable. In fact it can also be shown that $c^2 - 4ab > 0$, by substituting $D = \mu_m \tilde{s}/(K_s + s)$ from equation (4.6) into the expression for c, so that the roots have no imaginary parts and hence there cannot be any overshoot with damped oscillations in the behaviour of the chemostat. Experimentally, however, this is found to be possible, so that once again there is a discrepancy between theory and observation, requiring modification of the model. This has been discussed by Moser (1958).

If D is less than D_c, then the solution $\tilde{x} = 0$, $\tilde{s} = s_R$ can be shown to be unstable by substituting these values into equations (4.11) and (4.12). One then finds that the characteristic equation is $(\lambda - \alpha)(\lambda + D) = 0$, where $\alpha = \mu_m s_R/(K_s + s_R) - D$, so that there is a positive eigenvalue, giving instability. But if D is greater than D_c, then $\tilde{x} = 0$, $\tilde{s} = s_R$ will be stable. The behaviour of this system in relation to the parameter D is interesting in showing that stability is dependent upon certain performance limits, determined by the parameters.

Adaptive Behaviour of a Chemostat

The last observation is instructive in drawing attention to certain general principles of adaptive performance which we have already encountered in relation to the hierarchical relationships of the metabolic and the epigenetic systems discussed in Chapters 1 and 2 and which can now be formalized in relation to the chemostat. In equations (4.4) and (4.5) we have chosen to represent the state of the system in terms of two variables, x and s. The other terms are parameters, quantities which are either specified by the experimenter and held constant during an experiment, such as D and s_R; or quantities which are intrinsic to the type of cell being used, such as μ_m, K_s and Y. The latter are not under the control of the investigator except by choice of a different cell, but the former are. Suppose that the purpose of setting up a chemostat is for the collection of some biological product, say vitamins, so that the cells collected in vessel 3 constitute the pool from which material is drawn and processed according to demand. One could design the chemostat so that there was always a pool of some specified volume in vessel 3 despite varying demand by adjusting the flow rate, f, from vessel 1 according to the level of cells in 3. When this level drops below a reference value, L, a signal could go to the valve regulating f, opening it and so increasing the rate of production of cells; and vice versa. This signal would act as a negative feedback device, since if the level is too low, the rate of flow is increased and vice versa, exactly as feed-back inhibition operates in regulating the rate of flux of metabolites along a pathway to an end-product in the metabolic system. Evidently there is a fairly direct analogy here, although the control expressions will not in general be the same. If D in equations (4.4) and (4.5) is made a function of level in the pool, then another variable is being introduced into the system which was not considered in deriving these equations initially. Clearly there are many different types of expression which would satisfy the general requirement of a functional relationship between level, flow rate and demand, written in the form of differential equations. The chemostat would then become a homeostat, maintaining a relatively constant pool of product despite variations in demand, in the same way that a metabolic pathway maintains a relatively constant pool of end-product in relation to physiological demand.

However, if demand exceeds a value such that f results in $D = f/V > D_c$, then the system will clearly fail. How can it be designed to provide the flexibility or adaptability required for such eventualities? Another parameter which is under the designer's control is the volume of the chemostat, V. By increasing this, D is decreased for a fixed f, and so it can be kept less than D_c. To incorporate such variability, a new control circuit would have to be introduced which adjusted V (say by raising or lowering an adjustable overflow arm instead of having it fixed at a certain height) whenever the ratio f/V approaches the critical value D_c.

At this point it is useful to look at the relationship between $D\tilde{x}$, which is a measure of the output of the chemostat, as a function of D, in order to see what principles should govern the functional relationship between f and V. From equation (4.8) we see that

$$D\tilde{x} = DY\left\{s_R - K_s\left(\frac{D}{\mu_m - D}\right)\right\}.$$

This relationship gives the graph shown in Figure 4.2. Evidently there is a maximum which can be shown to occur at

$$D = \mu_m\left[1 - \left(1 - \frac{s_R}{K_s + s_R}\right)^{\frac{1}{2}}\right]$$

so that an optimally adaptive system would keep V at a value such that D remains in the neighbourhood of this peak as f varies. Changes in the quantities $\tilde{x}$ and $\tilde{s}$ following variation in V would be much slower than those in L, the level of product in vessel 3, Figure 4.1, as a result of changes in f, so there would be different relaxation times in response to these variations.

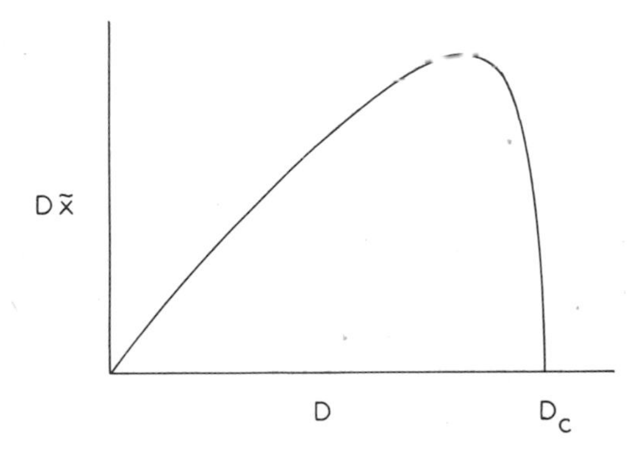

Figure 4.2. A graph showing the relationship between $D\tilde{x}$ and D for cells growing in a chemostat.

Changes in the capacity of the system as a result of varying V are analogous to the changes controlled by feedback repression, where the amounts of the enzymes are altered to achieve a change in flow capacity in a metabolic pathway. And once again we see the same principle operating: adaptability of performance is achieved by making parameters into variables, with defined functional dependence on other variables. Of course this always makes the system dynamically more complex. But this complexity can be held to a manageable level by employing the hierarchical principle which depends upon time scale: short term variations are handled by one level of control, longer and more extensive variations by a 'higher' level of control. The former we usually refer to as homeostatic and the latter as adaptive performance, although these are clearly relative terms. Analytically their definition arises from the relationship between variables and slowly changing ('secular') parameters and the relaxation times of each. Evidently there is no theoretical limit to the number of levels of adaptive operation which a system may have.

Cell Renewal Systems

The chemostat and its homeostatic modifications provides a simple model for certain types of physiological control processes operating at the level of cell populations. In the last chapter I discussed a possible theory of epidermal homeostasis, involving both temporal and spatial order which was achieved by the maintenance of a gradient of epidermal chalone across the tissue. There is evidence that chalones also act in the regulation of the rate of production of circulating cell populations in the blood, such as granulocytes (Rytomäa and Kivinieni, 1968), leucocytes (Garcia-Giralt *et al.*, 1970) and erythrocytes (Kivilaakso and Rytomäa, 1971). Let us see how one can account for the regulation of cell densities in these circulating populations by general control principles and compare this with the operation of the chemostat.

The first point to observe is that since there are different cell types in the blood and since each type can be regulated independently of the other, some specific means of control must be operating on the different populations. In theory this could be achieved by having separate growth compartments for the stem cells of each cell type and regulating the growth rate within each compartment by controlling the in-flow of a general nutrient such as glucose, as in the chemostat. However, the different cell types of the blood share the same physiological compartments in the organism which are liver, spleen and bone marrow. Within these tissues there is certainly histological heterogeneity, reflecting a partial compartmentalization of the different types of precursor or stem cells whose progeny differentiate into the mature cell types, but there is no evidence of separate circulatory systems for these different cells. Thus control must depend upon specific signals rather than the supply of a general nutrient, as occurs in the chemostat. The organism does have a degree of control over the nutrient supply to any tissue by regulating blood flow, so a valve such as operates in the chemostat still exists. However, this cannot achieve the fine discrimination necessary when the cell populations within the growth chamber are heterogeneous.

Such discrimination can occur if the differentiated cell types release a protein such as a chalone or other substance which specifically inhibits the growth rate of their precursor cells, as shown by Bateman (1974) and Bateman and Goodwin (1975) for blood leucocytes and erythrocytes in bone marrow cultures. These inhibitors may be regarded as feeding back from the circulatory blood, represented in Figure 4.1 by vessel 3, to the growth chamber, vessel 2, which now represents the blood-forming organs containing the different precursor cell types.

The growth rates of the precursor cells within the growth vessel would be functions of the physiological level of cells in the output vessel, measured by the concentration of the respective chalone, and of other specific signals which affect growth rate. For example, in the case of the erythrocyte precursor, the hormone erythropoietin plays an important role in stimulating division rate under conditions of physiological demand for greater oxygen carrying capacity of the blood, so the negative feed-back effect of the postulated erythrocytic chalone complements the action of this hormone, much as lactose and catabolite repressor operate together in regulating the overall activity of the *lac* operon. Evidently the simple principles of chemostat operation need to be extensively modified to model the behaviour of haemopoietic and other cell renewal systems. Flexible and versatile approaches to such modelling may be found in studies by Korn *et al.* (1973) and by Prothero and Tyler (1975). I would like to turn now to another aspect of cell population behaviour which is of considerable interest and importance, the phenomenon of cell synchrony.

Synchronization of Cells

In our study of the behaviour of population growth in the chemostat, no account was taken of the fact that this growth, at the level of the single cell, is a periodic process. As observed at the beginning of this chapter, the existence of a cycle in each member of a population makes possible the occurrence of cyclic behaviour at the population level. One situation which can result in a population cycle is the presence of an environmental periodicity which is sufficiently close to the mean generation times of the cells to force them into synchrony with it. An example of this is provided by the studies I carried out on the synchronization of *E. coli* growing in a chemostat to which a required macronutrient, phosphate, was added once every doubling time (Goodwin, 1969*a,b,c*). For cell growth in a chemostat, this doubling time is determined by the dilution rate, D, as one sees from equation (4.1), since at the steady state $\mu = D$. Growth being given by the expression

$$\frac{dx}{dt} = \mu x$$

which integrates to give the familiar exponential function

$$x = x_0 \, e^{\mu t},$$

it is evident that the value of t which results in a doubling of cell density is that value of T satisfying the relationship

$$2x_0 = x_0 \, e^{\mu T}$$

or

$$\mu = \frac{\log_e 2}{T}.$$

Thus $D = \log_e 2/T$ at the steady state, giving the relation between dilution rate and the doubling of population generation time, T. In my experiments bacteria were grown on medium lacking phosphate, with a flow rate requiring a mean doubling time of 90 minutes and a fixed amount of phosphate was added every 90 minutes. The behaviour of the culture was as shown in Figure 4.3, which shows the variation in cell density (A_{540}), cell number and

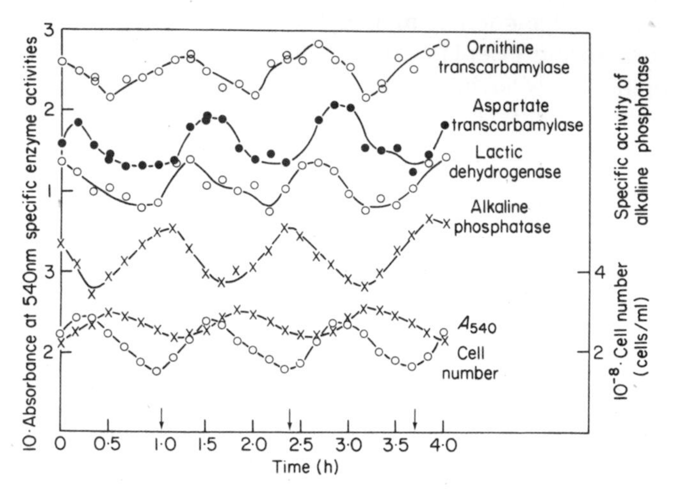

Figure 4.3. The behaviour of a variety of variables in a bacterial culture synchronized by periodic additions of the growth-limiting nutrient, phosphate.

the activity of a number of different enzymes. Under these conditions a relatively high degree of cell synchrony is achieved, of the order of 60%. The intrinsic variability of the division process in bacteria makes it difficult to achieve higher degrees of synchrony. Changing the environmental periodicity to a value different from the doubling time reduces the synchrony markedly, as one would expect.

The fact that cell divisions begin immediately after the pulse of phosphate is added to the culture suggests that phosphate is necessary for the completion of some events associated with cross wall formation and cell separation. Evidently the intracellular level of phosphate drops too low for these processes to be completed, although the bacteria continue to grow during this phase of phosphate deprivation, as seen from the growth curve shown in Figure 4.4. Here the chemostat variables have been converted to equivalent batch culture variables by the use of the transformation $X = x\,e^{Dt}$, X referring to a batch culture variable and x to that in the chemostat. The enzyme activity shown in this graph is aspartate transcarbamylase. The transformation of variables makes obvious the fact that there is no cessation of growth in the chemostat during the phosphate pulsing cycles. Evidently different processes have different sensitivities to phosphate level and in fact without such differences it would not be possible to synchronize cells by the use of a general nutrient such as phosphate. That bacterial physiology is

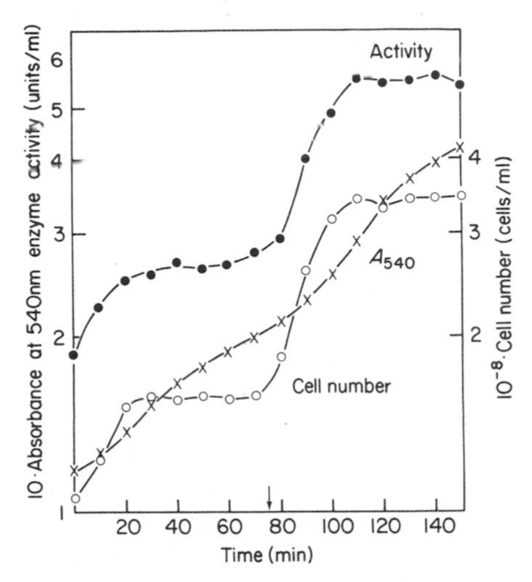

Figure 4.4. Chemostat variables transformed to equivalent batch culture variables to show division synchrony more clearly.

markedly periodic in all its aspects under these conditions is evident from the behaviour of the various enzymes plotted in Figure 4.3. Ways of exploiting this periodic behaviour to investigate dynamic aspects of bacterial physiology are given in the paper by Goodwin (1969*b*) in which the response characteristics of the *lac* operon to various perturbations are studied.

There are many examples of cell populations both of unicellular organisms and in metazoan tissues which show a population peridocity with respect to cell number, mitotic index or tritium index. However, it is not always easy to identify the immediate cause of the synchrony, since the situation is greatly complicated by the existence of cellular clocks which can themselves affect the cell cycle, as will be discussed shortly. Bacteria do not appear to possess such clocks, so the situation is simplified in this case. The metazoa certainly do have clocks but they are usually located in specialized organs or glands. Thus when one observes a 24-hour mitotic rhythm in, say, the epidermis or gastric epithelium of a mouse, it is likely to be produced by a 24-hour cycle in some hormone whose rhythm may be autonomous, as is the rhythmic activity of the adrenal gland in the hamster; or the hormonal periodicity may be induced by another periodic signal, possibly neural, which has an autonomous oscillation. Or it may be that the observed periodicity ultimately derives from an external period, such as a light/dark cycle, without any intermediate organismic clock activity. A great deal of research has gone into this intriguing area of biological behaviour which I will discuss in more detail later. At the moment I would like to consider briefly ways of approaching the analysis of the population periodicity under conditions of environmental forcing.

Again it is convenient to start with the general and simple model of cell growth in the chemostat and to particularize this behaviour to individual cases later. Since we are interested in population dynamics arising from individual cell behaviour, it is necessary to build the cell cycle into the model from the beginning. This can be done in the following manner. Let us first regard the cell cycle as a single compartment, whose traversal by a cell results in the production of two cells. Since, as discussed in the last chapter, there is considerable variability among cells passing around the cycle, we need a function which expresses the probability that a cell which divides at time τ will divide again in the small interval defined by $(t, t+\mathrm{d}t)$. Let this function be $f(t-\tau)$, so that the probability that a cell which divided at time τ will divide again in the interval $(t, t+\mathrm{d}t)$ is $f(t-\tau)\,\mathrm{d}t$. Let the number density of cells in a culture be $n(t)$. Then the number of cells 'born' during the small time interval in the past defined by $(\tau, \tau+\mathrm{d}\tau)$ may be written as $\mathrm{d}n(\tau)$. To find the number of cells dividing in the present small interval $(t, t+\mathrm{d}t)$, we must add up all the contributions from divisions occurring in the past, weighted by the probability distribution, $f(t-\tau)$ and multiplied by 2 since we assume that each cell gives rise to 2 daughter cells. This gives us the result

$$\mathrm{d}n(t)=\left[\sum_{r=1}^{m} f(t-\tau_r)\,\mathrm{d}n(\tau_r)\right]\mathrm{d}t$$

where r refers to the separate small time increments $\mathrm{d}\tau$, during which cells have been dividing in the past. Taking the limit as $m\rightarrow\infty$ so that the increments become infinitesimal, we get the integral

$$\frac{\mathrm{d}n(t)}{\mathrm{d}t}=2\int_{-\infty}^{t} f(t-\tau)\frac{\mathrm{d}n}{\mathrm{d}\tau}\,\mathrm{d}\tau$$

where the range of integration covers the whole history of the culture. If the experiment started at time 0, this can be written in the form

$$\frac{\mathrm{d}n(t)}{\mathrm{d}t}=2\int_{0}^{t} f(t-\tau)\frac{\mathrm{d}n}{\mathrm{d}\tau}\,\mathrm{d}\tau+\frac{\mathrm{d}n_0}{\mathrm{d}t} \qquad (4.13)$$

where $\mathrm{d}n_0/\mathrm{d}t$ refers to the initial conditions. This is known as the renewal equation. It can be used to describe the growth of cells in batch culture, or the rate of decay of synchrony as a function of the variability in cell doubling time, which is expressed by $f(t-\tau)$, the doubling time distribution function. Hirsch and Engleberg (1966) have made an interesting study of this problem, showing how one can obtain analytical solutions of equation (4.13) under certain conditions. Another model of decay of cell synchrony making use of a temperature function in relation to variability of generation times is given in a paper by Woolley and De Rocco (1973).

To extend equation (4.13) to a steady-state system with cell growth and cell removal as in the chemostat or in a living tissue such as the blood, it is necessary to introduce terms expressing the rate of passage of cells out of the growth vessel or stem cell population. Cell washout from a vessel with dilution rate D is a Poisson process, since it depends upon some random event. The probability of a cell surviving in the growth vessel for a period of time t is given by e^{-Dt}, so that the joint probability of a cell 'born' at time τ being still present during some later time interval $(t, t+dt)$ and dividing is

$$e^{-D(t-\tau)}f(t-\tau)\,dt \qquad (t>\tau).$$

There is also a washout rate $-Dn$, which arises in exactly the same way as the term $-Dx$ in equation (4.1). Thus the analogue of equation (4.13) for cell growth in a chemostat is

$$\frac{dn}{dt}=2\int_0^t e^{-D(t-\tau)}f(t-\tau)\frac{dn}{d\tau}\,d\tau-Dn+\left(\frac{dn}{dt}\right)_0. \tag{4.14}$$

Evidently if $D=0$ we get equation (4.13).

The next step is to compartmentalize the system and to introduce the environmental periodicity so that it affects the transition probability through one of the compartments. This can be done in a variety of ways. The following procedure was used by Dr M. Le Witt (1972) in a study of this problem. The cell cycle was divided into three compartments called W, U and V. Compartment W contains a number density y_W of cells which are waiting to progress through the rest of the cell cycle. The waiting time was assumed to depend upon the concentration of the limiting substrate, s, in the chemostat, and it was this substrate which was added periodically in Le Witt's computer simulation study. The U and V compartments contain y_U and y_V cells per unit volume, respectively, so that $y_W+y_U+y_V=n$, the total number of cells per unit volume in the chemostat. Passage of cells through compartments U, taken to consist of S-phase, G_2 and M, and V, which is part of G_1, was governed by probability distribution functions $f_U(t-\tau)$ and $f_V(t-\tau)$ of the same type as those in equation (4.13). These two compartments were distinguished in order to allow for flexibility in the model and computation of exit from U (i.e., cell division rate), but for many purposes they could be lumped together and described by a single transition probability function, $f(t-\tau)$.

Le Witt assumed that compartment W was a control point in the cell cycle corresponding to the G_1/S boundary, and that passage through it was regulated by an essential amino acid which determined the rate of synthesis of an initiator protein for DNA synthesis. This was modelled on the observation that arginine deprivation of Chinese hamster cells resulted in the arrest of the culture at some point in G_1 before entry into S-phase (Freed and Schatz, 1969). However, it is evident that the model may be interpreted in many different ways. Le Witt observed that a relatively high degree of cell synchrony could be obtained when the amino acid concentration was made to undergo oscillations with a period equal to the generation time, as shown in Figure 4.5. In this Figure can be seen the decay of synchrony of an initially

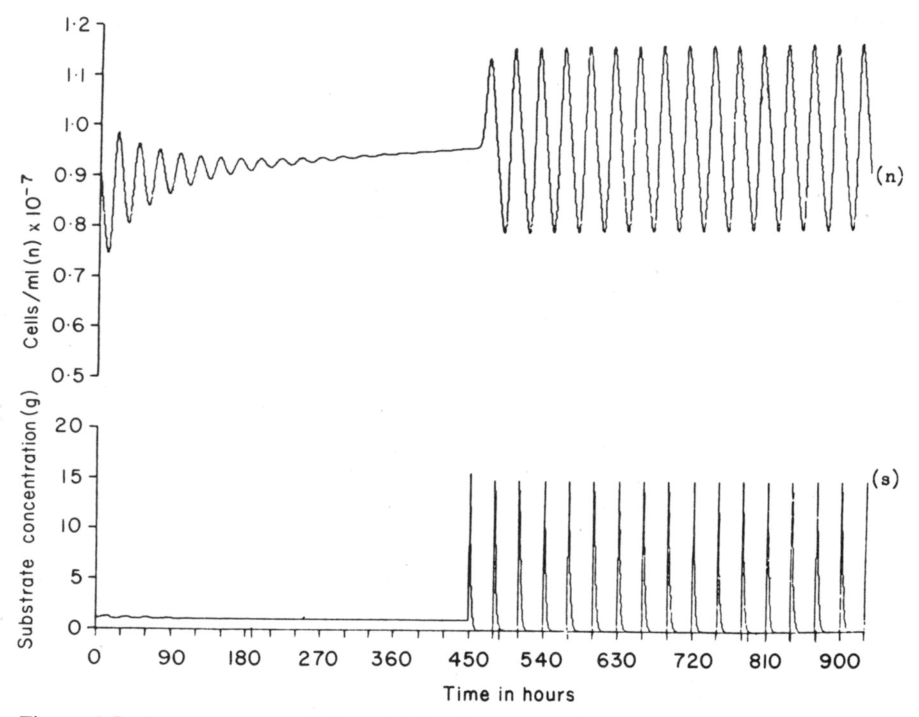

Figure 4.5. A computer simulation of LS cell synchronization in chemostat by periodic addition of a nutrient controlling *S*-phase entry.

synchronized culture placed in a constant environment and then the appearance of synchrony as a result of the oscillations in the amino acid concentrations. The synchrony in this system is 55%, comparing well with the value obtained by Goodwin (1969) experimentally with bacterial cultures in chemostats. Le Witt used parameter values for amino acid uptake rates

($K_s = 5{\cdot}7 \times 10^{-5}$ M), generation time (30 hours) and coefficient of variance of generation time (0.137) appropriate to LS cells (of mouse connective tissue origin), so his positive results on the induction of synchrony in the model confirm the expectation that cells can be synchronized by this means.

The effect of a decreased amount of noise or variability in the cell cycle, measured by the coefficient of variance (Le Witt used a transition probability with a Gaussian distribution), is shown in Figure 4.6 (coefficient of variance

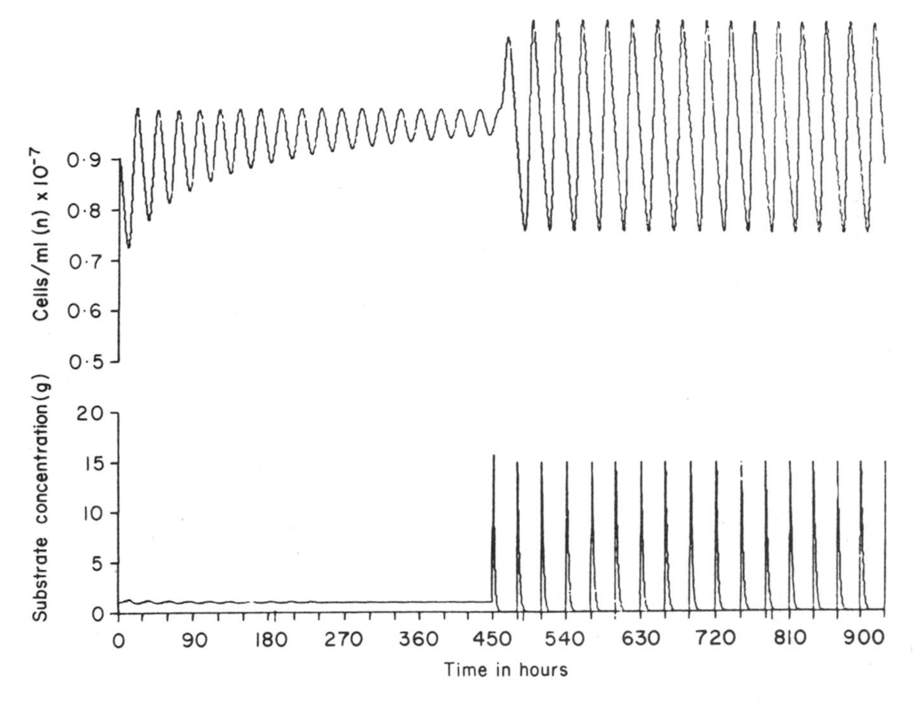

Figure 4.6. The effect of a decrease in the coefficient of variance of the cell doubling time distribution on the degree of synchrony of a simulated LS cell culture in chemostat.

0·1 instead of 0·137). Here one can see the slower decay of initial synchrony and an increased level of synchronization after the amino acid pulsing begins (63·7% synchrony). Changing other parameters such as s_R, the concentration of limiting amino acid in the in-flowing medium, the sensitivity of initiator protein synthesis to this limiting amino acid, and the half-life of initiator protein had little effect on the degree of synchrony attained over quite wide ranges. However, these parameters did affect the steady state of the cell population in the chemostat, as is shown in Figure 4.7. Here the half-life of initiator protein has been decreased to 2·5 hours from the standard value of 3 hours. The mean level of initiator protein in cells is therefore reduced, cells pass more slowly into the phase of DNA synthesis and the steady-state level of the population in the chemostat decreases. However, the degree of synchrony declines to only 54·6% from 55%. Le Witt explored many other interesting relationships of this kind, and his work strongly supports the expectation that mammalian cells in culture can be synchronized by this method. A rather different treatment of this problem using an alternative set of assumptions has been presented by Franck (1970).

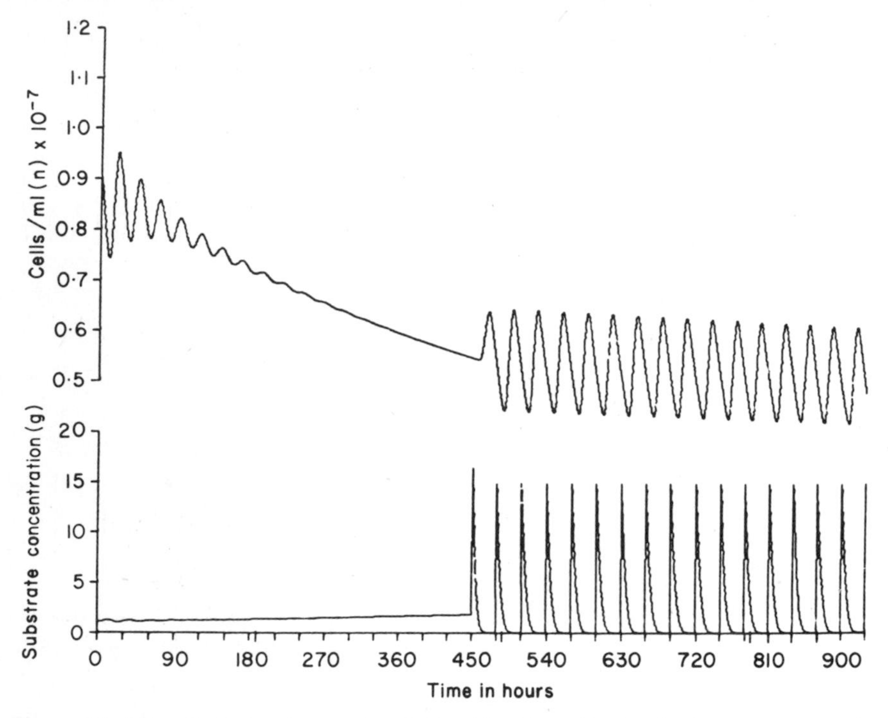

Figure 4.7. The effect of a decreased half-life of DNA replication initiator protein on the steady-state and condition of synchrony of a simulated LS cell culture in chemostat.

Diurnal Rhythms of Cell Division in Mammalian Tissues

As mentioned earlier, mammalian tissues show distinct 24-hour periodicities of mitotic and tritium index, among them the epidermis. It was in the course of studying this phenomenon in the mouse that Bullough and Lawrence (1960, 1961) were led to their investigations on chalones. They observed that the peak of mitotic activity in the epidermis of the mouse ear occurred during periods of rest, whereas when the mice were awake and active the mitotic index was low, the difference being about fourfold. A detailed study of the effect of adrenalin *in vivo* and *in vitro* and the effects of adrenalectomy on this rhythm confirmed that this stress hormone played a significant role in inducing the mitotic variation. Adrenalectomized mice showed a persistently elevated mitotic index and no diurnal rhythm, while mice subjected to stress or cells cultured *in vitro* and exposed to adrenalin showed depressed indices. Thus adrenalin evidently inhibits the rate of entry of cells into mitosis. Since the generation time of basal cells in the ear epidermis is three to four days, and the effects of altered adrenalin levels on mitotic index can be observed within a few hours, it would appear that adrenalin is affecting events fairly close to mitotis. However, other investigators such as Pilgrim, Erb and Maurer (1963), Pilgrim (1967), Brown and Berry (1968) showed that there was also a distinct rhythm of tritium index in tissues in some of which the G_1 phase alone is several days (e.g. stratified basal cells of the oesophagus). It therefore seems unlikely that the hormonal signal or signals, whatever their nature, were acting exclusively on premitotic events, since a long and variable G_1 would result in a distinct difference in the amplitude of the mitotic and tritium index rhythms due to the decay of synchrony in passage through a noisy compartment, G_1, contrary to observation. Thus there may be different rhythmic physiological signals acting upon different parts of the cycle to generate the observed rhythms. One signal with a rhythm of 24 hours acting on different phases of the cell cycle would tend to produce biphasic behaviour in a population, giving two peaks of activity corresponding to the two sub-populations each of which is captured by the signal at a different phase, unless the two phases were close to 24 hours apart. Thus if adrenalin acts towards the end of G_2 and also towards the end of G_1, inhibiting events at both phases, and if $S+G_2=24$ hours, then a diurnal mitotic rhythm would result. However, $S+G_2$ is usually considerably shorter than 24 hours (see Pilgrim *et al.*, 1963; Bruce, 1965). Thus it is not clear what the exact physiological origins of these mitotic and DNA synthesis rhythms may be. Some tissues, furthermore, show no mitotic or tritium index rhythm, so presumably they are insensitive to oscillating hormone levels. That mammalian physiology is organized on a 24-hour basis is clear from the work of many investigators (Halberg, 1963; Richter, 1965), but the adaptive significance of periodic behaviour in the cell cycle activity of different tissues is not at all clear. An interesting discussion of the relationship between the cell cycle and environmental rhythms from a comparative and evolutionary viewpoint is to be found in Bruce (1965). What we can conclude from the last two sections is that the cyclic nature of the growth and division process in cells makes them amenable to varying degrees of synchronization by periodic environmental signals, and that this type of behaviour can account in general terms for the diurnal rhythms of mitotic and DNA synthetic activity observed in mammalian tissues. There is no evidence that the cells under these circumstances are maintaining a state of population synchrony by means of interactions between themselves. Rather, the evidence is that whatever degree of synchrony is observed, it arises from an external, timing periodicity. One might anticipate, however, that under conditions of strong interaction between cells there would be a tendency for synchrony to develop, in the same way that within a syncytium such as in *Physarum polycephalum*, in an early insect embryo, or in fused cells mitotic synchrony arises. That this is not necessarily the case will now be shown.

Contact Inhibition

It has been known for a number of years that as the density of normal cells in a culture increases, their growth rate slows down until, when the culture is confluent with each cell completely surrounded by other cells and in contact with them, growth stops. This phenomenon of contact inhibition is of considerable physiological interest as an aspect of growth control. But since stem line cells such as the basal cells of the epidermis are also in contact with one another and yet continue to divide, the phenomenon does not seem at first sight to provide any insight into a possible regulatory process alternative to or complementary with the simple type of negative feed-back mechanism discussed in the last chapter. An additional observation of Lowenstein's (1967) adds an important dimension to this problem, however. He reported that many types of normal cells communicate with one another via tight junctions which form between the membranes of cells in contact and allow the passage of molecules of molecular weight up to about 5000. What would one anticipate to be the behaviour of a coupled population of cells which can transmit signals controlling their passage around the cell cycle, using the model that has emerged so far of a rather noisy cycle with an appreciable stochastic or probabilistic element causing considerable variability in the generation time? Will the coupling result in a more uniform periodic behaviour, so that the cells pass more regularly around their cycle; or will the variability increase by a summation process? And what may we anticipate about the amplitude of the control signals? We will find, of course, that the answers to these questions depend upon the particular assumptions we make regarding the stability characteristics of the dynamical processes underlying the cell cycle. But it is interesting to find that coupling can result in less organization of the whole rather than more, with the result that the control signals become so noisy and with such a small mean amplitude that they cease to function any longer as periodic regulators. The coupled population is then overwhelmed by its cumulative noise and the cell cycle effectively disappears.

This interesting possibility was suggested by Burton (1971) and explored in some detail in a recent paper (Burton and Canham, 1973). The essential assumption underlying this model is that the cycle in a single cell is controlled by an oscillation which can be represented by a Fourier series of sine terms and the interactions between cells can be represented by linear terms. Thus the dynamics of the whole coupled system is linear and different solutions superpose (i.e. add together). With the postulate that different cells have different frequencies, distributed in a Gaussian manner, it follows immediately that the coupled system will show very noisy behaviour indeed, and that the mean amplitude of any cell will be much reduced over its free-running amplitude, since out-of-phase oscillations will be in opposition. Burton and Canham show that in an asymptotic calculation for a very large number of cells, each interacting with n others with coupling constant K, the amplitude of oscillation in any single cell is reduced by the factor $(1/n)/K$. There will of course be the occasional large amplitude fluctuation, the control signal then exceeding threshold so that the cycle can proceed, but this is infrequent unless particular assumptions are made. The control signal can clearly be either of inhibitor or initiator type, since each functions by a threshold mechanism of some kind which would fail to operate if the amplitude is sufficiently decreased.

This result is not restricted to the model proposed by Burton and Canham. The feedback repression oscillator described in Chapter 2, with dynamics given by equation (2.5) can be used to represent the behaviour of the control signal for initiation or repression, as suggested by Smith and Martin (1973). I discussed the behaviour of coupled oscillators of this kind in my book (Goodwin, 1963). Assuming symmetrical coupling as would arise from diffusion so that $k_{12}=k_{21}$ in the expression equation (24), page 109 of *Temporal Organization in Cells*, for the talandic energy of the coupled oscillators, one gets for the ratio of the mean positive amplitudes of the coupled to the uncoupled system the expression

$$\frac{(A_+)^c_{x_1}}{(A_+)_{x_1}}=\frac{\pi}{2}\frac{|H|^{\frac{1}{2}}/h_{12}}{[\sqrt{(h_{11}h_{22})}/h_{12}+1]\tan^{-1}|H|^{\frac{1}{2}}/h_{12}}<1 \tag{4.16}$$

where

$$0<h_{12}<\sqrt{(h_{11}h_{22})}, \qquad \text{and} \qquad |H|=h_{11}h_{22}-h_{12}^2.$$

Thus we see that coupling non-linear oscillators of this kind also results in a reduced amplitude of oscillation, and it can be shown that the more oscillators are coupled, the greater is this reduction. In this model it is not necessary to introduce a Gaussian distribution of oscillator frequencies, for a basic postulate of the theory is the occurrence of noise in the system. This enters into the statistical mechanics in the form of a system parameter called the talandic temperature ('oscillatory' temperature). If the coupled cells are identical in all respects other than their instantaneous dynamic state, then one would take $h_{11} = h_{22}$ in equation (4.16). Clearly this has no effect on the result.

The feature of both these models which results in this decreased amplitude of coupled oscillators is the fact that both have the same kind of weak stability. They are conservative systems with orbital stability, to use the technical term. If a stronger type of stability in the oscillators had been assumed, such as asymptotic orbital stability characteristic of limit cycles, a different result would be anticipated, since these tend to become entrained or synchronized (Winfree, 1967). Presumably this is why mitotic synchrony occurs in *Physarum*, for example.

This observation takes us back to the original problem of contact inhibition and growth control. Evidently different types of behaviour may result from cellular interactions occurring via tight junctions, depending upon the strength of the interaction, the stability characteristics of intracellular oscillators and the noise level of molecular processes within cells. The range of possibilities is of interest because once again it makes available to cells a variety of consequences depending upon their state. There is also a degree of freedom in relation to whether or not tight junctions will form when cells are in contact. Some types of tumour cell, for example, fail to establish such communication channels (Lowenstein and Kanno, 1967) while others do not become contact inhibited even though they have tight junctions (Azurnia and Lowenstein, 1971; Borek *et al.*, 1969). Again small groups of cells may become more active when in contact with one another, an observation made by Cone (1969) and called contact facilitation. All these cases could be explained by a general model involving an intracellular oscillation with particular dynamics, noise level and some form of intracellular coupling. Chalones could act in concert with such a mechanism, possibly acting at the cell surface to increase cell affinities and the incidence of tight junctions, as well as or instead of the proposed effect on the transition probabilities for different parts of the cycle.

Cellular Clocks

The cycle of cell growth and division, though a periodic process, does not in fact have the characteristics of a clock. This is not because of its variability under constant conditions, which would simply make it an unreliable chronometer. It is because the mean period of the cell cycle varies with environmental factors such as composition of the medium, illumination and temperature, the Q_{10} being greater than 1. This of course is exactly what one expects of a process whose rate is dependent upon metabolic activities which themselves vary with these influences. But since it is the function of a clock to measure defined intervals of time irrespective of environmental variations to which it may be exposed, it should be compensated against these; i.e., it should be a temporal homeostat, keeping constant time. The problem of designing reliable chronometers was a considerable challenge to the ingenuity and craftsmanship of 16th and 17th century inventors and artisans, because these were essential to the development of navigational accuracy in this era of world exploration. Independence of environmental variation was achieved by isolating the chronometer from disturbances through mechanical means: enclosure in a waterproof casing, suspension in gimbals to prevent mechanical shock, etc. Temperature compensation was more difficult to achieve and it was not until the latter half of the 18th century that the Frenchman, Le Roy, designed compensation balances so that moments of inertia of fly-wheels were not altered by changes of temperature. It is interesting to note in passing that the Polynesian islanders, who had to navigate over distances of 2000 miles or more in their archipelago, could do so as well as Captain Cook, although they had no mechanical chronometers or sextants. Evidently they used their own clocks, together with a variety of highly-developed cognitive skills learned as navigational apprentices. The intriguing problem remains how this ability and the even more dramatic navigating abilities of migrating birds is achieved, and what the nature is of the biological process which gives this freedom in space by virtue of a firm grasp of time. The relationship between maps and clocks is a very basic one in biology, as we will see in the next chapter in relation to embryogenesis; but at the moment our concern is primarily with chronometry.

The phenomenon of an approximately 24-hour rhythm of activity of various kinds in organisms occurring independently of any obvious environmental periodicity has been known for many years. Such rhythms are now referred to as circadian (*circa diem*, about one day) and they are the most intensively studied manifestation of internal time-keeping within organisms, although other periods of time such as tidal cycles, weekly, monthly (lunar), seasonal, annual and longer cycles have been reported as elements of the chronometry of organisms. I will restrict myself here to circadian rhythms and of these only examples which allow a detailed study of the phenomenon at the cellular level.

One of the most interesting organisms to have been studied in relation to its circadian behaviour is the unicellular marine dinoflagellate *Gonyaulax polyedra*. This micro-organism is one of those responsible for the luminescence of the oceans, the luminescent activity rising to a peak in the middle of the dark phase, from 11–12 p.m. Cultures of *Gonyaulax* can be maintained in the laboratory, allowing a detailed study of their circadian physiology. An investigation by Sweeney and Hastings (1958) showed that not only is the flashing luminescent rhythm periodic, but there is also a rhythm of luminescent glow with peak at about 2 hours before the end of the dark phase and a rhythm of cell division with a maximum incidence of dividing cells one half an hour later than the glow peak. These rhythms persist for weeks when cultures are placed in constant dim light (100 ft candles) after a period of exposure to alternating light (900 ft candles) and dark, 12 hours of each (abbreviated to LD 12:12). The period of the rhythm under constant conditions is no longer exactly 24 hours, as it is under LD 12:12, but is close to one day. If the temperature is altered, the period of the rhythm changes slightly. For example at 18 °C the period is 22·9 hours, whereas at 25 °C it is 24·7 hours. Thus the rhythm has a Q_{10} less than 1; i.e. the clock is not simply compensated, but over-compensated, in its response to temperature! However, the main point is that the period stays in the neighbourhood of 24 hours despite variations in temperature, unlike the generation time.

Under an LD 12:12 regime at 21·5 °C, the culture of *Gonyaulax polyedra* studied by Sweeney and Hastings (1958) had a mean generation time of 1·5 days. Since the great majority of cell divisions in any day take place just before dawn, it is evident that individual cells generally undergo alternate generation times of 1 and 2 days to give a population average of 1·5 days. (They could in fact have other combinations of generation times, such as n 24-hour generation times and m $24p$-hour generation times in $n+pm$ days, with $(n+m)/(n+pm)=\frac{2}{3}$, where n, m and p are integers, but solutions other than $n=m=1, p=2$, are relatively unlikely.) What is of particular interest in the present context is that the cell cycle in this organism is under the control of an endogenous or intracellular clock, which forces or 'gates' the cell cycle by some means in the same way as the forcing of synchrony between cycling cells and an external periodicity. Cells with mean generation times other than the period of the driving oscillation will be inhibited from entering division during their out-of-phase cycles (assuming inhibitory control), but general growth will continue during this inhibition, so that they can complete the next cycle more quickly than average. Thus there is no difference in the working of the synchronizing process in these two cases except for the origin of the synchronizing signal, this being intracellular in the case of circadian clocks.

However, there is a very intriguing problem that arises at this point and that is why cultures of *Gonyaulax polyedra* in continuous dim light after being synchronized by LD 12:12 remain in synchrony for weeks, without any sign of decline in the amplitude of the rhythm after the second cycle. If cultures are maintained in continuous bright light, the rhythm decays rapidly and the culture becomes totally asynchronous, cells dividing at any time of day or night. It would appear that conditions of dim light allow intercellular communication of some kind so that the cells remain in phase throughout their cycles, but the nature of this interaction is unknown. Similar observations have been made by Edmunds (1971) on cell division rhythms in *Euglena gracilis.*

The Physiology of Cellular Clocks

An intensive period of research on the nature of cellular clocks coincided with the emergence of molecular biology in the late fifties and throughout the sixties as the dominant mode of analytical thinking in biology, and much of this research was conditioned by molecular concepts. A natural candidate for the cellular oscillator thought to underlie the clock was one of the negative feed-back circuits which had been found to play such a central role in cellular physiology. However, the relative temperature-independence of the clock suggested that there must be something in addition to a control loop whose individual steps were regulated by metabolic processes such as macromolecular synthesis and enzyme catalysis, with Q_{10} values between 1 and 2. A suggested candidate for this temperature-insensitive aspect of the clock was diffusion (Ehret and Trucco, 1967). Bacteria are too small for diffusion to be a significant aspect of control, but fortunately they are ruled out of play by the fact that they appear not to have circadian rhythms. So the overall hypothesis that tended to be entertained was that macromolecules such as mRNA were synthesized within the eukaryotic nucleus, these were then released into the cytoplasm and diffused to sites of interaction with ribosomes, whereupon proteins were synthesized and themselves diffused to functional sites, inducing changes of cytoplasmic state. These cytoplasmic changes were then communicated to the nucleus and exerted a negative feed-back effect, again by diffusion, thus completing the cycle. The time periods of these processes could plausibly be made to fit the required 24-hour time scale (see Goodwin, 1963).

The obvious way of testing this sort of model was to interfere with one or other of the steps in the sequence of events in an attempt to stop the clock. Actinomycin D, puromycin or actidione, and various enzyme inhibitors were used in a great variety of organisms with circadian rhythms to arrest mRNA synthesis, protein synthesis and enzyme catalysis, respectively, in an attempt to probe into the workings of the postulated molecular oscillators. The results of many such experiments are reported in symposium volumes on biological clocks such as the *Cold Spring Harbor Symposium* (1960), *Circadian Clocks* (Aschoff, 1965), the *Cellular Aspects of Biorhythms* (Mayersbach, 1969), etc. Although many very interesting results were reported, no clear cut picture emerged. And towards the end of the sixties it became evident that the physiological origins of the biological clock were almost as little understood as they had been a decade earlier. This situation is well described in Bunning's (1967) book *The Physiological Clock* which reads rather like a detective story, each avenue of approach being systematically explored in terms of molecular biological paradigms and each one leading to an inconclusive result, the clock constantly eluding the investigator.

The essential difficulty in this area of research is characteristic of that in many other biological fields. This is the problem of identifying a primary rather than a secondary or later aspect of the process one is seeking to understand and analyse. Karakashian and Hastings (1962) showed that at a concentration of 0·2 μg/ml, actinomycin D suppresses the luminescence rhythm in *Gonyaulax*. Since growth was only partly interfered with under these conditions, it is evident that some DNA-dependent biosynthetic events relating to periodic bioluminescent activity are more sensitive to the antibiotic than others connected with growth. In the same paper it was shown that a similar result is obtained with the use of puromycin to stop protein synthesis. These observations certainly suggest that the synthesis of mRNA and protein are necessary conditions for the occurrence of periodic bioluminescent behaviour in *Gonyaulax*. Studies with other organisms showed that respiratory inhibitors would interfere with circadian periodicities, showing that the rhythms require an energy source. However, it was frequently found that when organisms which had been exposed to a drug for some period of time were returned to normal conditions, the physiological periodicity started up again at the same phase as it would have had if left undisturbed. It does look as if the antimetabolite was interfering not with the physiological rhythm underlying clock function but with the expression of some physiological process which is driven by the clock, a point that repeatedly emerges in Bunning's (1967) analysis. The clock appears to remain unaffected by the treatment.

A crucial test of a primary effect on the clock itself would be to change its time, its phase. Until very recently, the only antimetabolite which had been found to do this was arsenite, which caused a phase delay in the *Gonyaulax* glow rhythm (Hastings, 1960). However, this provided little insight into the mechanism of the clock. What one requires is a chemical (or chemicals) whose action mimics the phase-shifting effect of light, and whose cellular target is known so that some inference can be made about the underlying physiological process involved in biological rhythms. Such a substance has now been found and with it the study of biological clocks has entered a new era. But before it had been found, research in this field had led to a set of apparent paradoxes whose only resolution was to thoroughly abandon the original set of molecular biological paradigms which had so heavily conditioned thinking in this field and to take a new approach. This turns out in fact to be a very classical one, physiologically speaking. The story of the reaching of the apparent impasse in terms of molecular biological paradoxes is as follows.

The role of the nucleus in cellular circadian rhythms can be investigated in a very direct manner in the case of the unicellular alga *Acetabularia*. This remarkable cell consists of essentially three parts: the rhizoid which houses the nucleus and anchors the plant to its substratum; a circular cap about 1 cm in diameter; and a narrow stalk several centimetres long which connects these. The cytoplasm is continuous throughout, consisting of a thin cylinder in the stalk surrounding a central vacuole. It undergoes active cyclosis, mixing the contents of all parts of the cytoplasm which contain chloroplasts, mitochondria, ribosomes and various other inclusions. The plant is enclosed in a continuous cell wall. In the next chapter I will be discussing some of the regenerative and morphogenetic properties of this organism.

The very large size of *Acetabularia* makes it possible to measure physiological activities of single cells and to carry out grafting experiments which would be virtually impossible with other cell types. The nucleus is easily removed simply by cutting off the rhizoid and a different nucleus can be put in its place by grafting another rhizoid to the stalk, a technique developed by Hämmerling (1934). The existence of a circadian rhythm of photosynthetic activity in this organism was established by the work of Sweeney and Haxo (1961) and by Schweiger *et al.* (1964). The latter investigators showed that the rhythm persisted in the absence of the nucleus, while Sweeney *et al.* (1967) demonstrated that periodic photosynthetic activity continued in enucleate plants in the presence of actinomycin D, puromycin and chloramphenicol. Thus neither a nucleus nor protein synthesis is necessary for the circadian clock to function in *Acetabularia*. However, if nuclei are transferred either with rhizoids or as isolated organelles between plants which have been kept 180° out of phase with respect to their LD schedules, then after several days in continuous dim light the phase of the rhythm is determined by that of the nucleus. Furthermore, the addition of actinomycin D to plants with nuclei resulted in the disappearance of the circadian rhythm of photosynthetic activity after several days. Evidently the nucleus does exert an influence upon the behaviour of the cytoplasm; but equally obviously, the cytoplasm does not need a nucleus to show circadian behaviour.

The chemical which provided the clue to the solution of this puzzle is the ionophore, valinomycin, which increases the rate of entry of K^+ into cells by interaction with the membrane. It was found by Bünning and Moser (1972) that this substance was very effective in imitating the phase changes caused by light on leaf movements in *Phaseolus*. This observation immediately turned attention to the role of the membrane and ion concentration changes as the possible basis of physiological rhythms. Recent studies by Njus, Sulzman and Hastings (1974) and by Sweeney (1974*a*) have confirmed and extended these observations with respect to circadian rhythms in *Gonyaulax*, and membrane models have been advanced by these investigators which provide plausible descriptions of the basic clock mechanism.

One such model is the following, advanced by Sweeney (1974*b*) in explanation of the *Acetabularia* paradoxes. An important observation regarding membrane behaviour in the chloroplasts of *Acetabularia* is that they change their volume in a diurnal manner, suggesting a circadian rhythm of permeability. Sweeney postulates that the nucleus also undergoes rhythmic changes of state associated with membrane permeability changes and that it is a cycle of exchange of substances between these organelles and the rest of the cytoplasm, determined by organelle permeability, that underlies the circadian clock of *Acetabularia*. There are many ways of realizing such a rhythm or oscillation, but Sweeney proposes that there are two metabolites of relatively low molecular weight, X and Y (not macromolecules), X being induced only in the cytoplasm by Y, while Y is induced in organelles by X. The permeability of the organelles is controlled by the concentration of X within them, and both substances are subject to decay. Since we have distinguished between the molecules in these two compartments, let us call them X_c, Y_c (cytoplasm) and X_O, Y_O (organelles). Then Sweeney's model can be represented by the following equations

$$\frac{dX_c}{dt} = aY_c - bX_c + f(X_O)(X_O - X_c)$$

$$\frac{dY_c}{dt} = g(X_O)(Y_O - Y_c) - cY_c$$

$$\frac{dX_O}{dt} = f(X_O)(X_c - X_O) - \alpha X_O$$

$$\frac{dY_O}{dt} = \beta X_O + g(X_O)(Y_c - Y_O) - \gamma Y_O.$$

Here the expressions such as $f(X_O)(X_O - X_c)$ and $g(X_O)(Y_O - Y_c)$ represent the rate of movement of X and Y across the organelle membrane, whose permeability is given by the functions f and g of X_O. What gives these equations the possibility of an oscillatory solution is the existence of these permeability functions, which could be quite non-linear. Then if we were to plot the concentration of X_O against Y_O, for example, we could find the following sort of behaviour. Calling the point O phase 0, the concentration of X_O is then at a maximum and the organelle membranes are at their least permeable. Y_O is being synthesized within the organelles but X_O is decreasing because of decay. As X_O decays the membrane becomes more permeable to Y_O, which then escapes and reduces the level within the organelle. X_O will then also decline rapidly, but as Y in the cytoplasm, now Y_c, induces X_c there will be a slow build up of X_O within the organelle and the cycle starts again. A complementary cycle of X_c and Y_c will take place.

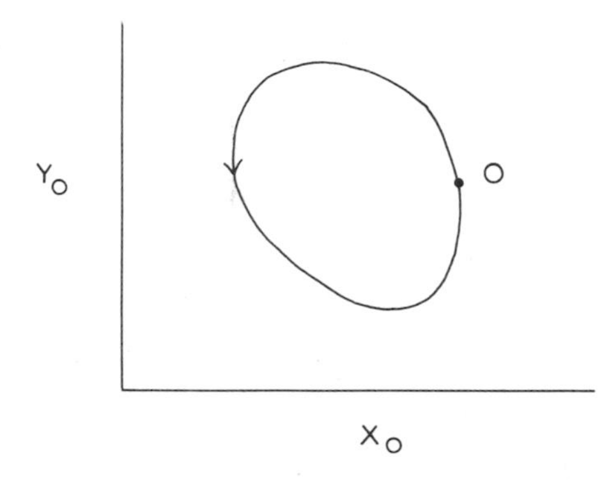

Figure 4.8. Phase plot of the variables X_O and Y_O.

This model could account for the observation that the circadian rhythm of photosynthetic activity in *Acetabularia* is independent of a nucleus, of mRNA and of protein synthesis, providing that the enzymes and precursors for X and Y are present. It can also explain how a large organelle such as the nucleus in a particular phase of the cycle could phase shift the cytoplasm. But it does not account for the decay of the circadian rhythm in entire plants exposed to actinomycin D. As Sweeney points out, this requires an additional observation, which is that in the presence of a nucleus, cytoplasmic proteins become unstable, whereas they are stable in its absence. The phase-shifting effects of light and temperature on the circadian clock are explained by the sensitivity of organelle membrane permeability to these influences.

An interesting clock model which emphasizes the importance of metabolic compartments as essential features of circadian time-keeping has been advanced by Cummings (1975, 1976). He uses a simple metabolic sequence with a positive feed-back loop, which as we have seen produces dynamic instabilities. The metabolic species of this sequence are located in a cytoplasmic compartment (the cell membrane or cortex) which equilibrates slowly with the rest of the cell. It is this slow equilibration which results in a long period of oscillation, of circadian magnitude. Cummings' (1976) equations have the following form:

$$\dot{x} = A - k_1x - S(y)x$$
$$\dot{y} = S(y)x - k_2y,$$

where $S(y)$ is a sigmoid function describing the positive feed-back process whereby y accelerates its own synthesis, while k_1 and k_2 are the equilibrium constants between the compartment and the cytoplasm. Cummings identifies x as cAMP and y as 5′AMP, thus focussing attention on the role of these metabolites in the circadian clock mechanism. Using a linear approximation to $S(y)$ in the neighbourhood of the steady state, defined by $\bar{x}$, $\bar{y}$, Cummings shows that the frequency of the oscillator is

$$= [\,\bar{y}/\bar{x}(k_2^2 - k_2k_1) - k_1^2]^{\frac{1}{2}}.$$

By choosing the difference between k_1 and k_2 to be sufficiently small, this can readily give a frequency of one per day. Another interesting feature of this oscillator is the very slow rate at which its amplitude increases as the trajectories wind out towards the limit cycle from inside, a point to which we shall return later in discussing some experimental observations on clock behaviour. This model has many interesting experimental implications, especially those relating to cAMP and 5′AMP compartmentalization. The reader is referred also to Edmunds and Cirillo (1974) for a thoughtful discussion of clock physiology.

What about the phenomenon of temperature compensation in biological clocks? An interesting explanation of this has been proposed in a recent paper by Njus *et al.* (1974). They advance a fairly detailed molecular realization of a membrane model in which changes in the positions and orientations of membrane proteins, as visualized in the fluid mosaic model of Singer and Nicolson (1972), are responsible for changes in the ion transport activity of the membrane. Ion concentrations in the non-membrane phase are themselves taken to regulate these changes of activity and distribution of membrane proteins, so one has again the basic mechanism of the self-regulating negative feed-back system operating. Since these membrane proteins are relatively stable to degradation, one has here a model which is insensitive to inhibitors of macromolecular synthesis and to enzyme inhibitors. The circadian physiology of the cell is conceived to be regulated by ion changes, a proposition which has become widely accepted in its general form in view of the large number of hormones which appear to act primarily on membranes and so transmit their primary effects by ionic signals. Photoreceptors in the membrane acting as K^+ gates explain light sensitivity and the remarkable rapidity with which light signals can act in causing phase shifts.

Temperature compensation is considered to result from lipid adaptation, a phenomenon which occurs in all poikilotherms and involves changes in the lipid composition of the membrane to maintain constant fluid quality over a wide range of temperature (e.g. Barenska and Wlodaver, 1969). Njus *et al.* suppose that protein migration and co-operative interactions between macomoleccules in the membrane are basic kinetic aspects of the clock mechanism, so constancy of fluid properties would render the process relatively insensitive to temperature except at extreme values, where clock function is in any case impaired, along with many other physiological processes. This model has many interesting experimental consequences, as well as providing a very plausible molecular formulation for physiological time-keeping.

It thus appears that the elusive biological clock is being tracked down and that a primary component is the membrane. The models advanced by Sweeney (1974), Njus *et al.* (1974), and Cummings (1975) are conceptually simple and elegant and are founded upon physiological principles which are independent of the molecular biology of gene control. The relative isolation of the clock mechanism from the general physiology of the cell in the theories described above is a result of compartmentation and the location of processes on membranes, together with an independence of macromolecular synthesis. What is emerging in this field is a study of kinetic mechanisms with long relaxation times arising from slow diffusion or equilibration processes, extending greatly the dynamic range of the metabolic system by including the properties of its phase boundaries. This inclusion of membrane behaviour in intracellular dynamics will be as important in the study of spatial order as in the current analysis of temporal order.

THE DYNAMICS OF THE *DROSOPHILA* ECLOSION CLOCK

In terminating this chapter I would like to consider one of the classic examples of a metazoan clock in order to illustrate the imaginative use of some very powerful but simple analytical concepts. The organism is *Drosophila pseudo-obscura* and the circadian rhythm is that of the emergence of adult flies from pupae, known as eclosion. This normally occurs in the early hours of the morning just before dawn. A population of *Drosophila* larvae can be synchronized simply by transferring them simultaneously from continuous light (LL) to continuous dark (DD). After pupation these larvae will emerge as adult flies, a group undergoing eclosion every 24 hours, approximately, over a period of days. This rhythm is thus observed in a population, since each individual fly emerges only once.

However, a great deal of imaginative work has established that this popula-tion behaviour can be traced to a clock within each fly which gates eclosion in the same way that the circadian clock of *Gonyaulax* gates or regulates the permitted time of cell division (Pittendrigh, 1960, 1965, 1966; Engelmann, 1966; Chandrashekaran, 1967). Once a fly is ready to emerge, it is permitted to do so only at a particular time of day. Evidently we have here an interaction between a periodic process, the clock, and the developmental process, which is progressive.

Studies on *Drosophila* cannot answer the same sort of questions as those posed above in relation to the molecular processes which might be involved in circadian behaviour such as the role of the nucleus, different molecular species, membranes, etc. However, one can ask some searching questions about the stability characteristics of the clock and about its dimensionality. Notice that so far the minimum number of variables assumed to be necessary for an oscillation in a biological system is two, as in Figure 4.8 of this chapter, or Figure 2.8 in Chapter 2. However, there is an even simpler type of oscillation than this, dimensionally-speaking, and that is what has been referred to as a generalized relaxation oscillation. This is a periodic function of time which can have different periods but practically no change of amplitude. The state of such an oscillator can be defined in terms of one variable only, which is its phase; i.e. it is like an ordinary clock which can be set to any time and will then run at fixed speed, with the only visible indicator of state being the position of the hands. These show only time or phase, measured between 0 and 24 hours or 0 and 2π radians, respectively. From very extensive and ingenious studies of the behaviour of biological clocks in response to various types of perturbation, Pittendrigh and Bruce (1957) proposed a model in which the chronometer was assumed to be an oscillator of this relaxation type whose phase could be instantaneously reset by a light or other signal, but whose amplitude remained unaffected. Any transcience in the observable response to the signal was then regarded as occurring in that process which is controlled by the clock and by which its existence is made manifest. This model is widely used and has had very considerable success.

I mentioned above that an important property of biological clocks is their responsiveness to particular environmental signals which act as reliable indicators of solar time, the most important of these being light. This is simply an example of the phenomenon of synchronization, discussed above. The effect of a synchronizing signal on a periodic process will evidently depend upon the phase of that periodic process, since in general the signal could act either by advancing or by retarding phase; i.e. set the clock forward or backward. A lot of the experimental work on biological clocks has been devoted to an exploration of this behaviour and many phase response curves have been determined which give a plot of the phase shift ($\Delta\phi$) induced by a given signal applied at a particular time or phase of the biological clock. In the case of *Drosophila*, this might be 15 minutes of white fluorescent light given to a population of pupae at a particular time after they were transferred from LL to DD to set their clocks in synchrony. When compared to a control culture, this population will start to undergo eclosion either earlier or later or at the same time, the phase response being then either positive or negative or zero, respectively. The period of the eclosion rhythm is unaffected by this treatment. A typical curve is shown in Figure 4.9. Observe that there is a point of sharp discontinuity in the curve where a phase delay approaching 12 hours is suddenly converted into a phase advance when the light signal is given at about 18.00 hours on the clock. Instead of the system being set back any further than 12 hours, it is suddenly advanced by the light signal. What does this mean in terms of the operation of the clock? At first sight it appears to suggest that there is a threshold-effect in its operation such that after about 18 hours the clock can be set forward by a light signal, but before this it will be set back. However, Winfree (1970, 1971) has provided an alternative explanation for this discontinuity and in so doing has initiated a new way of thinking about the underlying constraints which govern periodic behaviour in organisms.

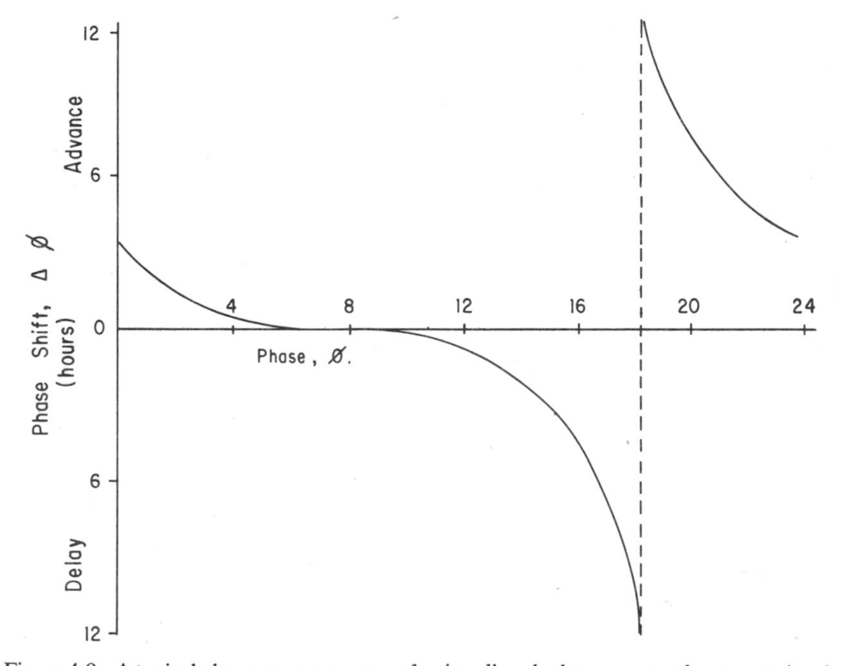

Figure 4.9. A typical phase response curve of a circadian clock to a strong phase reset signal.

The essence of Winfree's argument derives from purely qualitative, topological arguments about the possible behaviour of biochemical or physiological oscillators and the judicious use of Occam's razor to keep the picture as simple as possible. Recognizing first that a one-dimensional relaxation oscillator of the type described above which has constant amplitude and variable phase is necessarily an idealization, he explored the qualitative behaviour expected of the simplest plausible type of biological oscillator, which is one with two variables, amplitude and phase (or, equivalently, two variables X and Y). This oscillator is extinguished by continuous light in *Drosophila pseudo-obscura*, so the dynamics of the system changes from those of a limit cycle, with trajectories converging on a stable closed cycle, to those of a system with point stability, as shown in Figure 4.10. The limit cycle is labelled C in this Figure, and the point O is the singularity within the cycle from which the trajectories unfold to the cycle. It is the (unstable) steady state of the system in the dark. The point P is the stable steady state of the system in the light, there being then no oscillation. A light signal will drive any point on the limit cycle towards P, and depending upon where the initial point is on the limit cycle and how strong the signal is, it would end up somewhere within the limit cycle and displaced towards P at the end of the signal. It will then travel out towards C, after the light is switched off, and where it joins up with C compared with the control point

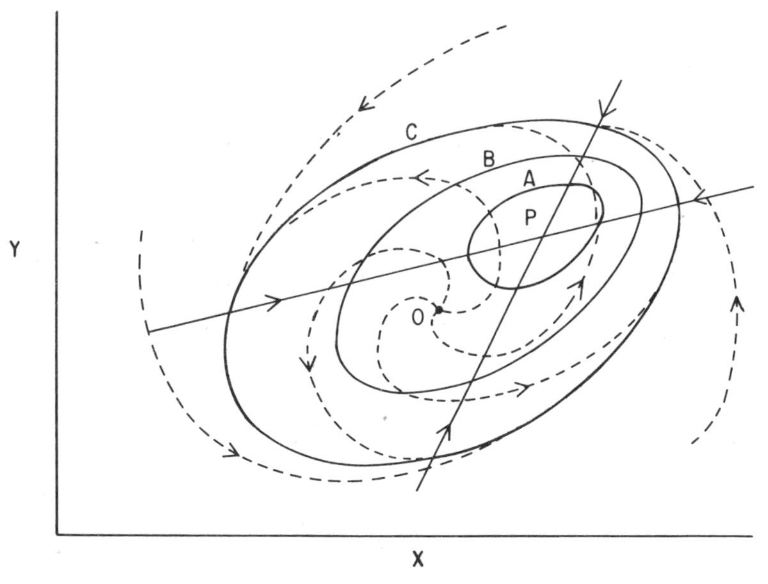

Figure 4.10. A dynamical explanation of the origin of the discontinuity in the phase response curve of a circadian clock in terms of the topological properties of a limit cycle.

travelling steadily around *C* determines the extent of the phase shift, and whether positive or negative (advance or delay). How, then, do we get a curve of the type shown in Figure 4.9 with a discontinuity?

Winfree argued as follows. Suppose that initially we have a population of flies whose clocks are distributed around the limit cycle, *C*, so that each fly has its clock functioning but they are all at different phases. Now if a light signal is given, the points on *C* representing the states of the clocks will be driven towards *P*, and at the end of the light signal they will be distributed on some curve such as *B* in Figure 4.10. The points will then move out again towards the limit cycle, rejoining it after some period of time dependent upon how far *B* is from *C*. Each point will experience a phase shift compared with its control point, either forward or backward depending upon the dynamics of the return to the limit cycle (roughly speaking, the length of the return trajectory). But observe that these points will be distributed again over the whole of the limit cycle and will preserve their original neighbourhood relationships. This is because each spiral arm from the singularity, *O*, cuts the curve *B* in one and only one point.

If, however, a stronger light stimulus is given so that the points representing the clocks are driven to the closed curve *A*, closer to *P* and such that all points are on the *P*-side of *O*, then the situation is quite different. Now each spiral trajectory winding out from *O* to *C* cuts the curve *A* in either two points or none. The trajectories which do not intersect *A* will define an empty section of the final distribution of points on *C*, while points which were originally on that part of *C* corresponding to this empty section after perturbation will return to points on the opposite part of the cycle. In the middle of this set there is a point of discontinuity where a sudden change occurs from a 12-hour advance to a 12-hour delay. This analysis shows that strong and weak perturbations of biological clocks should lead to quite different phase response curves, only the former showing a discontinuity. Winfree deduced that this must be so, and then discovered from the literature that indeed weak light signals produce phase shifts without discontinuity. Furthermore, there are no other types that have been observed, which strengthens the assumption that the dynamics may be represented qualitatively as in Figure 4.10.

With this initial success, Winfree went on to exploit this kind of topological thinking. He reasoned that if a point such as *O* exists, it should be possible to find that light signal which, when applied to pupae at a particular phase of the cycle, will drive them to *O*. This being a steady state, the clock will there be stopped and the population should be timeless, arhythmic. Since nothing was known about the quantitative aspects of the dynamics, looking for a critical signal or stimulus of this kind was like searching for a needle in a haystack. The remarkable thing is that Winfree (1970, 1971) found it by dint of some very clever experimental design that allowed him to scan thousands of *Drosophila* and to identify the critical stimulus, which was a 50-second pulse of white light at 6·8 hours circadian phase. This seemed to establish the correctness of the dynamical picture proposed. However, there was a problem: instead of winding out again to the limit cycle, the clocks appeared to be stopped dead at O. This could be explained on the assumption that the return from O to the limit cycle is very slow, and eclosion takes place before the clocks are functioning again.

Winfree (1973) has in fact shown that it is possible to affect the amplitude of the eclosion clock by light stimuli such that the oscillator continues to go around its cycle but on a smaller trajectory, with little change in its period. This requires a modification of the phase picture of Figure 4.10. One could, as mentioned above, draw the trajectories winding out slowly from the singularity, O, in accordance with Cummings' model (1976), which shows precisely this behaviour. This type of explanation has also been suggested by Aldridge and Pavlidis (1976). However, Winfree (1975) has suggested an alternative explanation of his observations which is based upon the concept that the organism has a whole population of clocks and tells the time according to their degree of collective synchrony. Light stimuli can disrupt this synchrony and give a poorly entrained population with a low mean amplitude. This introduces a new level of complication into the time-keeping story, one which requires new forms of analysis. Analytical models for this type of collective behaviour of rhythmic populations will be considered in Chapter 7. In concluding this chapter, let me just observe that the beauty of the approach to biological oscillations adopted and exploited by Winfree is the way in which he has demonstrated that very general qualitative analysis can guide experimental study in an extremely useful and powerful manner.

Chapter 5

MORPHOGENESIS: APERIODIC ORDER IN ONE DIMENSION

INTRODUCTION

I SHALL consider in this chapter the generation of one-dimensional axial order in organisms and its relationship to the processes we have been considering thus far. Having started our enquiry in the timeless, spaceless world of homeostasis where the primary preoccupation is stability in homogeneous systems and having discovered that the mechanisms of negative feed-back and allosteric behaviour used to achieve stability and discrimination create the potential for oscillatory behaviour and hence temporal organization, we come now to the question of how spatial structure emerges from non-linear kinetics and temporal order. This has already been touched upon briefly in Chapter 3 in relation to structural homeostasis in the epidermis. It was found that the temporal order of the differentiation process, when associated with cell displacement from the basal layer of the epidermis, provided a model for the stability of tissue structure to such disturbances as wounding or transient carcinogen application. This global stability resulted from a graded spatial distribution of a molecular control substance, the epidermal chalone. This protein was assumed to affect the state of differentiation of cells as well as the rate of passage of the cells around the cycle, in virtue of a reciprocal interaction between the processes of cell cycling and cell differentiation (see Figure 3.8). In the present chapter I will implicitly assume that any molecular species which has a graded spatial distribution over a developing system and is responsible for initiating changes of shape and form, hence is acting as a morphogen, exerts its influence by some means similar to that proposed for chalone action. Thus the morphogen is assumed to affect either directly or indirectly the molecular control machinery of cells and to influence the rates of synthesis of various substances which relate to the differentiation process. There is no discontinuity of action in moving from problems of homeostasis and stability to those of temporal organization, and finally to those of temporal and spatial order. The hierarchical principles used throughout this analysis continue to operate, so that higher-level systems, so called, continue to include in their operation all relevant activities of the lower-order ones. I shall continue to assume the importance of molecular processes while considering the mechanisms whereby ordered heterogeneity of molecular activity is achieved over developing and regenerating systems.

We can start on the analysis of spatial order by extracting certain rather general principles from the model of epidermal homeostasis. The stability of the spatial organization of the tissue as represented in this model depended upon a differential distribution of rate of production of the diffusible molecular species, epidermal chalone, and on a boundary which acts as a sink, the basal membrane, with the circulatory system beneath. The upper boundary of the mature tissue consists of impermeable keratinized squames which block the escape of chalone and hence results in its greater accumulation in this region. A section through the skin from surface to basal layer showing the expected distribution of chalone concentration would look roughly as shown in Figure 5.1. In this model each cell is a source of chalone

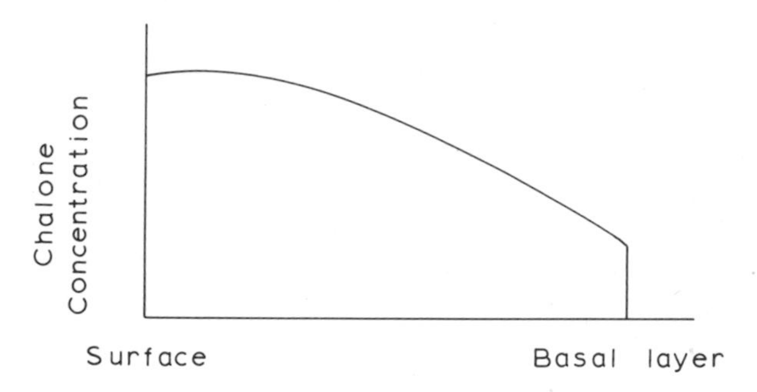

Fig. 5.1. A qualitative picture of the G1 chalone concentration in the epidermis.

and may also be a sink, if the protein is degraded. In addition, cells migrate from basal layer towards the surface. The exact shape of the gradient will be determined by the rates of production and destruction of chalone in the cell in different regions, the permeability of cells to it, its rate of diffusion, its rate of escape across the basement membrane and the rate of cell movement distally. The system is one in dynamic equilibrium. The details are not, of course, available, so we cannot set out to find an exact solution. However, we can consider the general class of system to which this model belongs as a particular example and discuss the evidence from different experimental systems for variants within the class. All of these involve metabolic gradients, diffusion and cell movement as important features of the space-ordering process. It thus becomes necessary to alter the focus of our enquiry from problems centred about stability in homogeneous systems to problems of dynamic fields and their spatial stability.

Polarity and Axial Gradients

The model of epidermal homeostasis involved a gradient of a substance along the axis of organization of the tissue which is perpendicular to the basement membrane, as shown in Figure 5.1. The tissue may thus be said to be polarized with respect to this axis of spatial organization: proximal cells (adjacent to the basement membrane) are in a different state from distal cells (at the epidermal surface) with a graded distribution of cell state in between. In somewhat more abstract terms, a polar axis results from the existence of a reference point or surface defining the origin of an axis and a direction for measurement along it. What is being measured and how this information affects cell state is, in most developmental systems, a major research problem. But polarity of some kind is an almost universal feature of both multicellular and unicellular organisms. I shall consider the nature of this order in some primitive metazoa which have been used extensively as research material because they embody one-dimensional axial organization in idealized simplicity; yet the essential basis of their polar stability remains the subject of continuing study. These are the hydroids, *Hydra* and *Tubularia*. *Hydra* has an extremely simple overall structure, being essentially a tube with a mouth (hypostome) surrounded by a ring of tentacles at the distal end and an adhesive foot (basal disc) at the other. Its tissue structure is also very simple, consisting of two epithelial layers known as the ectodermis (outside) and the gastrodermis (inside), both attached to a thin membranous layer containing collagen-like material, the mesoglea. Its overall dimensions are about 1 to 2 cm in length, depending upon the species, and about 1 mm wide when the organism is extended. A photograph of *Hydra littoralis* is shown in Figure 5.2.

These organisms have remarkable capacities for regeneration of a whole from small parts and reconstitution from fused fragments. If a section is cut from the central body part, a new head (hypostome and tentacles) will regenerate distally and a new base proximally, preserving the original polarity of the tissue. This polarity can be reversed by treatments such as an electrical field [demonstrated by Lund (1921) on *Obelia* and by Rose (1963) on *Tubularia*] or grafting a hypostome proximally and removing it several days later (Wilby and Webster, 1970). If the section is very short, however, then heads may be produced at both ends, or polarity may be reversed, so that there are conditions of instability with respect to the preservation of original axial order in *Tubularia* and *Obelia*.

A great deal of early research into the metabolic origins of polarity in hydroids was done by C. M. Child who developed the concept of metabolic gradients, particularly of oxidative metabolism, as the underlying cause of

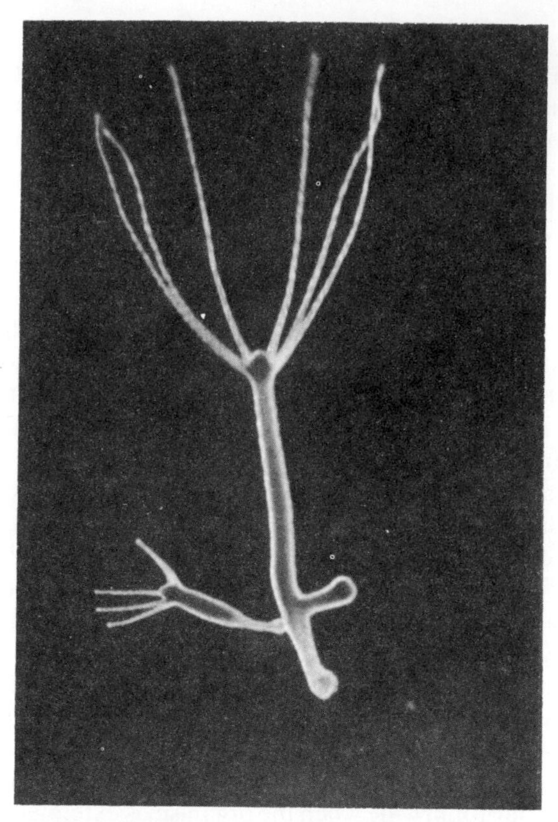

Figure 5.2. A photograph of *Hydra littoralis* showing tentacles and hypostome, gastric region, budding zone with one growing bud and one differentiated *Hydra* ready to detach from the parent, and the basal disc. (Courtesy of Dr Amata Hornbruch).

axial order. His studies with inhibitors and dyes suggested that metabolic activity is greatest at the hypostome and decreases towards the basal disc. The theory of metabolic gradients elaborated by Child (1941) has been extremely fruitful in guiding both research and theory-construction, despite contradictory evidence about the existence of a gradient of oxidative metabolism. In their general form, when linked with ideas of diffusion gradients (which require metabolic gradients of some kind), electrical fields and intracellular communication, Child's ideas remain as a foundation for much thinking about the dynamic nature of polar order in developing systems. A lucid review of the developments in this field is given in Webster (1971).

As it is always the head which regenerates first in mid-gastric sections of hydroids, with the other parts differentiating later, it would appear that the most distal part acts as an organizer for the rest of the system and so defines the origin of the polar axis. Over a wide range of size of regenerate, the approximate sizes of parts such as the hypostome, tentacles, digestive zone and peduncle remain the same, so that the organism has the capacity not only to regenerate its parts but to scale them in proper proportion to one another, that is, to regulate its parts. However, with very small regenerates this property of scaling fails in *Tubularia* and *Corymorpha* and in these the head parts are appreciably larger in proportion to the others. The general strategy of regeneration is to produce a functional head first and then for the other parts to be formed out of the remaining tissue, with some degree of interaction giving rise to scaling or regulation. Thus the hypostome acts as a dominant or organizing region, in accordance with Child's postulates. The question that then arises is the nature of this organizing activity. Some investigators have assumed that there may be a hypostomal inducer produced distally and that a gradient of this substance then determines the region of head formation. There has been some evidence for such a substance, believed to be a polypeptide (Lentz, 1965; Schaller, 1973). It could act rather in the manner of a chalone with respect to its proposed role in inducing cell differentiation in the epidermis. However, the stability of axial order in the epidermis depended upon two other factors: the occurrence of a localized growth zone in the basal layer and the existence of a chalone sink across the basement membrane. In *Hydra* there is no evidence of a local growth zone (Campbell, 1967); nor is there positive evidence that the basal disc acts as a sink for a hypostomal inducer. However, it is clear that hypostomes and peduncles are in many respects antagonistic, since a graft of a peduncle to the distal part of the regenerating section will prevent a hypostome from regenerating there and will cause polarity reversal. This could be either because the peduncle is acting as a sink (i.e., destroying the hypostomal inducer); or because it is producing an antagonist, an inactivator or inhibitor of the hypostomal inducer; or finally, because it is producing its own inducer. Evidence for the latter possibility together with an interesting discussion of these alternatives is given by Newman (1974). We must consider now various approaches to an understanding of this kind of axial order.

The recent development of analytical models in this area has been greatly influenced by the ideas and experimental observations of Wolpert and his group, who have worked primarily with *Hydra* and the chick. A general analysis of the problem presented by the phenomenon of pattern formation and morphogenesis in developing systems and an elucidation of the essential concepts required for their resolution was presented by Wolpert (1969). Basic to this analysis was the concept of positional information, and the interesting postulate that whatever the physiological processes might be that are involved in generating spatially distributed information (e.g. a gradient of a particular substance), they are likely to be universal throughout the animal, and possibly also the plant, kingdoms. Wolpert argued that in regulative systems such as the sea urchin or the amphibian embryo, the establishment of an axis in relation to which cells differentiate according to their position requires (i) a reference point from which measurement begins; (ii) a direction for measurement along this axis: and (iii) a scale-adjustment mechanism which can alter the units of measurement along the axis so that the total number of units remains invariant to changes of axial length. In Wolpert's terminology these are (i) a reference cell or region, termed α_0; (ii) polarity, which is the direction in which positional information is measured; and (iii) a second reference cell or region, termed α'_0, at the opposite end of the axis from α_0. The values of the gradient at α_0 and α'_0 are assumed to be fixed by some means which is independent of the length of the axis so that a regulating gradient results. Given such an axis, a cell located at some point along it will have available to it information which is specific for that position, and can then differentiate according to some 'code' for interpreting the positional information. The distinction between positional information and its interpretation is a very important one in Wolpert's analysis and it makes the assumption of universality plausible. Different species could employ different codes for interpreting positional information provided by axes which are the same or similar throughout the phyla, thus achieving variety of phenotype while conserving a basic epigenetic mechanism. However, one must not place too great a burden on the process of interpretation, since to do this is simply to transfer a problem from one area to another. Wolpert gets perilously close to this when he argues (1972) that the concept of positional information allows one to dispense with the notion that there is any necessary connection between the structure of a morphogenetic field or prepattern and the spatial order that becomes manifest when the field has been expressed. Thus, he argues, the same (linear) gradient could produce a pattern such as the French Flag in one organism and the Stars and Stripes in another. This is very economical on mechanisms of positional information, but extremely expensive, I should imagine, on interpretative machinery. Waddington's (1972) position is that the quest for universality of morphogenetic mechanisms is probably illusory and that a diversity of distinctly different processes is at work in developing organisms. As will emerge in this and the next two chapters, my view is essentially universalist, like Wolpert's but I see the organism working more as an analogue than as a digital computer and hence do not share his belief in the capacity of developing systems to interpret a monotonic gradient as a mosaic, a periodic pattern, a structure with circular symmetry, or any other form, the only constraint being on the informational dimensionality of the field and the resulting structure.

Wolpert's analysis of pattern formation was deliberately carried out in abstract terms, to avoid any prejudice towards one model or another. It is this neutrality and generality of presentation which has stimulated the search for models satisfying the basic requirements described and the experimental exploration of their consequences. Later in this chapter I shall consider a model of morphogenetic field formation which does away with the last of Wolpert's requirements, since it achieves regulation without either a scale-adjustment mechanism or a second reference point in the system, thus simplifying the basic axiomatics for setting up a field of positional information. The price paid for this simplification is that regulation does not occur completely along an axis which is reduced to less than some critical length. However, this is precisely what is observed in organisms such as *Tubularia* and *Corymorpha*, so it would appear that requirement (iii) goes beyond what is demanded by experimental observation. More recent studies from Wolpert's laboratory on the development of the chick limb have led to a quite different model from that arising out of regeneration studies in *Hydra*. This is a rather more dynamic one than that used in explanation of *Hydra* phenomenology, involving a growth process and a developmental timing mechanism, but the concept of positional information is retained at the price of some ambiguity of meaning. In the next chapter I shall describe this model and suggest an alternative which provides a unified view of both types of process.

It is useful at this point to consider a mathematical theory presented by Gierer and Meinhardt (1972) which describes the properties of a fairly wide class of process conforming generally to the behaviour of axially-ordered systems. These authors assumed that there are two different processes operating in such systems, whose interaction results in a graded distribution of cell state. One of these was postulated to have the property of short-range activation, characteristic of induction, whereby a positive influence is transmitted from one cell or group of cells to its neighbours, as occurs when sub-hypostomal tissue is transplanted to a mid-gastric region of *Hydra* and a hypostome is there induced, with local tissue contributing to the induced structures. It is short-range because only cells neighbouring the graft are affected. The other process they assumed to be long-range inhibition, as exemplified by the suppression of hypostome formation distally by the presence of a grafted hypostome proximally, the effective distance of the inhibitor being several hundred cells. The differences between these ranges of action are assumed to arise from different diffusion rates of specific activator and inhibitor substances, although there are other ways whereby short- and long-range interactions may occur, as will be discussed later. Gierer and Meinhardt also assume an initial graded distribution of source and/or sink densities for these substances which represent the initial polarity of the tissue. Furthermore, they suppose that the source and sink densities are distinguished from the concentrations of the substances themselves, and are governed by processes which have characteristically longer relaxation times than those for the activator and inhibitor, in accordance with the evidence obtained in hydroids. Thus the model applies to events occurring within a few (say 4 to 12) hours of any particular experimental perturbation, which is the time required for primary gradients to be set up. These postulates lead to differential equations of the general type shown below:

$$\left.\begin{aligned}\frac{\mathrm{d}a}{\mathrm{d}t}&=\rho(x)+f(a,h)-\mu a+D_a\frac{\partial^2 a}{\partial x^2}\\ \frac{\mathrm{d}h}{\mathrm{d}t}&=\rho'(x)+g(a,h)-\nu h+D_h\frac{\partial^2 h}{\partial x^2}\end{aligned}\right\}\qquad(5.1)$$

where x is the spatial variable along the axis. Here a is the activator, h is the inhibitor, $\rho(x)$ the source distribution of the activator, $\rho'(x)$ that of inhibitor, $f(a, h)$ and $g(a, h)$ represent the effects of a and h on their respective synthetic rates, μ and ν are first-order decay constants, while D_a and D_h are the respective diffusion constants, it being assumed that $D_h \gg D_a$. The types of function chosen for $f(a, h)$ and $g(a, h)$ depend upon the assumptions made concerning the kinetics involved, such as whether the interaction between a and h is of competitive or non-competitive type, and the stoichiometry of the reactions. In general these will be non-linear control functions describing the activity of enzymes with co-operative kinetics and allosteric behaviour of the type discussed in Chapter 1. It is the occurrence of these functions which gives the equation (5.1) the properties required for the generation of stable gradients from appropriate initial conditions on the source terms ρ and ρ'. This equation belongs to the general category of diffusion-reaction system, the first example of their use in the context of morphogenesis being a now-classical paper by A. M. Turing (1952) on the possibility of generating periodic waves of morphogen concentration over a tissue. Meinhardt and Gierer (1974) apply their model to the behaviour of *Hydra* and other developing systems, demonstrating that it satisfies many of the required properties. The question of the relationship between activator and inhibitor concentrations, acting as morphogens, and their effect on cell differentiation are, of course, left out of account in this treatment, the distinction between the two processes being made in accordance with the criterion of different relaxation times. As they point out, consistency does not allow any deductions about the validity of the model, only biochemical and physiological studies providing the evidence required to elucidate the details of the molecular mechanisms involved. The utility of the model is in showing that a variety of different types of mechanism can be made to fit the general scheme, thus providing a category of classification for a fairly diverse set of models. These apply not only to hydroids and flatworms, but to any developmental system with axial organization and polar order.

The major short-coming of this model is that it does not account for regulation, since if the linear dimensions of the system are altered one gets a gradient which fails to keep positional information invariant over the new length. This is shown in the first paper by Gierer and Meinhardt (1972). Before looking for a model with this property, let us turn to an organism whose developmental behaviour cannot be explained on the basis of static gradients of the type we have been considering up to now in relation to spatial ordering mechanisms. This will lead us to consider a rather different way of generating a gradient from anything we have discussed so far.

Pacemakers and Activity Waves

Among the vast diversity of organisms available to biologists for study, there are those which embody, in particularly simple form, patterns of organization which are regarded as the essence or archetype of certain developmental processes. The hydroids do so with respect to axial organization and regeneration in the metazoa; the true slime mould, *Physarum polycephalum*, does so with respect to nuclear synchrony and the mitotic cycle; the epidermis plays this role in relation to tissue homeostasis. The selection of ideal representations of processes which are in fact realized in a variety of forms always betrays the ideology or the prejudice of the selector, but so long as we are aware of this it is a useful procedure. We now come to an organism which embodies, for many developmental biologists, the ideal representation of the transition from spatial and temporal disorder to the four-dimensional organization characteristic of the developmental process. This is the cellular slime mould, the species *Dictyostelium discoideum* having received most attention. From it we learn about a type of communication between cells which we have not yet encountered, and which may play an important general role in morphogenesis.

In the cellular slime moulds the phase of growth is sharply distinguished from the phase of development and differentiation. So long as there is a plentiful supply of bacteria on which to feed, the mould consists of a population of myxamoebae which move about by means of pseudopodia which form at any part of the cell surface, ingesting bacteria as they move. These vegetative amoebae are therefore apolar. The cell generation time is about three hours, and each ameoba behaves quite independently of its neighbours. There is no evidence whatsoever of any spatial or temporal order in the population apart from that which might arise from the spatial distribution of food source or the geometry of the surface. When the food supply becomes exhausted, however, growth ceases, cells become less mobile and cell divisions occur much less frequently, although they are observed occasionally. After a period of about eight hours, known as interphase, a new form of behaviour begins to appear if the amoeba density is sufficiently high. Individual cells distributed at random throughout the population begin to release periodically a chemical signal, originally called acrasin, which propagates from cell to cell, since cells receiving the signal are themselves stimulated to release the chemical (Schaffer, 1962). Concomitant with the appearance of signalling among cells, polarity appears so that amoebae develop an anterior and a posterior end. This appears to be a necessary condition for effective chemotactic response to the signal. Thus signal reception by a cell results in a twofold response; release of acrasin, so that the signal is transmitted to neighbouring cells; and chemotactic response, whereby cells travel towards the signal origin by amoeboid movement.

Acrasin has now been identified as cyclic 3′-5′AMP (Konijn *et al.*, 1968), which we have already encountered as the glucose distress signal of carbon-starved bacteria (Chapter 1). It would seem that starving amoebae of *D. discoideum* have similar metabolic circuitry which they have exploited to elaborate a signal-transmission system between cells under conditions of food deprivation. The period of signal release from the individual cells varies between three and seven minutes, with a mean value of about five minutes. These are just the periods we would expect for oscillations in the metabolic system and an interesting model based upon the allosteric control principles described in Chapter 1 has been presented by Goldbeter (1975). There is apparently nothing genetically distinctive about these cells, there being simply a random distribution of cell states over the initially disorganized population.

An important aspect of signal transmission is that amoebae enter a refractory phase of unresponsiveness for a period of about two to seven minutes, depending upon developmental age, after having released a pulse of cAMP. Also cAMP is rapidly broken down by an extracellular phosphodiesterase, secreted by the amoebae (Bonner *et al.*, 1969). Thus the dynamics of signal propagation are just like those occurring in neurones where a neurotransmitter substance such as acetylcholine is released at nerve endings and is there rapidly destroyed by acetylcholinesterase, thus ensuring the possibility of the recurrence of another signalling event. The refractory period ensures one-way propagation of the signal and also establishes well-defined boundaries between pacemakers competing for amoebae, since no amoeba can simultaneously serve two masters. The dynamics of the competitive interaction between pacemakers is determined both by the refractory period of amoebae and by their relative frequencies, a faster pacemaker having the advantage. These interactions result in a fairly even partitioning of an initial amoeba population among a number of pacemakers, with well-defined boundaries of dominion between them (Cohen and Roberston, 1971*a*,*b*). Very striking time-lapse films of the process of aggregation have been made by several workers in this field, showing the waves of amoeboid movement spreading out from pacemakers and the establishment of boundaries between them. The waves can be either concentric circles or spirals, the dynamics of which have been described and analysed by Durston (1973).

During the process of aggregation, amoebae stream together making contact as they approach the pacemaker cell and then form a compact mass of cells which pile up into a heap with the pacemaker cell or region at the top, carried up by the inflowing amoebae (Gerisch, 1968). This mound topples over and becomes an elongated, sausage-shaped mass of cells, now all in intimate contact, which is known as a slug. Pulsatile contraction waves continue to travel from anterior to posterior parts of the slug and may be connected with its movement, which consists of a random migration for a period of a few hours. The slug then stops moving horizontally and transforms its movement into a vertical one, sending up a stalk and making at the top of the stalk a cone of spores. Periodic movement is evident throughout this development. The net result is a fruiting body consisting of stalk cells and spore cells, the proportions of these two types (1:2) being fairly constant over a considerable range of size. However, very small fruiting bodies have a disproportionately large number of spores, in the same way that very small hydroids have disproportionately large heads or hydranths. The result of the whole developmental process, which takes 30–40 hours in all, is the transformation of a disorganized population of anti-social amoebae into spatially organized, differentiated fruiting bodies.

The aspect of this process which draws our attention to an important morphogenetic mechanism not considered by static gradient models is the periodic signalling which occurs throughout development in *D. discoideum*. This is a particularly ingenious way of calling dispersed amoebae together. Since the signal is propagated actively by the cells, the distance over which it can travel is virtually unlimited, unlike the severe spatial restrictions that would be imposed if a source cell attracted neighbours by maintaining a static gradient of chemotactic substance. However, an alternative mechanism is for cells to signal continuously rather than periodically so that each cell acts both as a source and a sink, the result being a process rather similar to that described by the Gierer and Meinhardt (1972) model. This global approach to cell aggregation has been developed by Keller and Segel (1970) (see also Segel, 1972), who analyse the process in terms of a continuous field description rather than the microscopic view taken here and in the papers of Cohen and Robertson (1971, 1972). An interesting analysis of these alternative approaches to the problem of aggregation and morphology in the slime mould is given in a paper by Nanjundiah (1974). He shows that for *D. discoideum*, the values of the quantities relevant to the relative efficiency of the two alternatives (diffusion constant, size of pulse of cAMP, threshold level for response and period of signalling) are such as to make the pulsatile mechanism much more effective.

A system of units capable of periodic wave propagation of this type has the properties of an excitable medium. The question now arises whether developing systems are, in general, excitable in this sense and what the consequences of such behaviour might be with respect to their spatial and temporal organization. Could we elaborate a unitary view of morphogenesis by combining aspects of the behaviour of *D. discoideum* with those deduced from the behaviour of hydroids, to develop a general theory which could explain the origin of polar order and periodic phenomena as well as the construction, regulation and stabilization of gradients? This is obviously aiming at some degree of universality before tackling the difficult and complex problems of morphogenesis in higher organisms, and in particular the problems of neuroembryology. But before attempting to construct such a theory, there is a very important aspect of morphogenesis which we have to take into account. This is the fact that certain unicellular organisms are capable of undergoing morphogenetic transformations, regeneration and regulation in ways indistinguishable from those occurring in multicellulars. Furthermore, there are phases of development in the metazoa where morphogenesis occurs in the absence of cellular partitions, such as the early stages of insect development. Thus if we want to attain a minimal level of universality for our morphogenetic theory, it must be one which does not depend upon interactions between cells. Just as we discovered that an understanding of the biological clock is to be found in the processes occurring within single cells, so now we shall see that the basis of the morphogenetic process may be found in intracellular organization. Multicellularity then becomes a strategy for increasing the size of organisms and for achieving fine detail of differentiation within tissues and organs, not for morphogenesis itself.

Morphogenesis Within Single Cells

The organisms that have contributed most to the evidence that gradients and fields of essentially the same kind as those operating in the metazoa are responsible for spatial organization within single cells are the green marine algae referred to collectively as the Siphonales, and the ciliate protozoa. It is satisfying to a universalist view of the type we are now seeking that the distinction between plants and animals becomes irrelevant in this context, as it has already in the juxtaposition of hydroids and the Acrasinae in our discussion of multicellular morphogenesis and as in the study of biological clocks in the last chapter. The alga that has provided most insight into the type of model required for an understanding of spatial order in single cells is, once again, *Acetabularia*. And just as we found that this organism has guided our attention to membrane-control mechanisms as likely candidates for biological clock generators, we will now find that it directs our focus to membranes as the structures responsible for the initial generation of form. So time and space measurement become closely related to the properties of biological surfaces.

The organizational characteristics which make *Acetabularia* ideal for the study of morphogenetic mechanisms are essentially its large size, its powers of regeneration and the ease with which the nucleus can be removed or grafted from one species with characteristic morphology to another. In addition, the clear differentiation of the cell into cap, stalk and rhizoid can be exploited for the investigation of morphogenetic determinants (see Figure 5.3). The early work which established *Acetabularia* as an extremely important experimental organism was mainly that of Hämmerling (1934, 1963). He showed not only that the stalk regenerates a cap after amputation, but that regeneration occurs after the rhizoid containing the nucleus has been cut off. This regeneration can take place many weeks after removal of the nucleus, the time of regeneration being controlled by light. (No regeneration occurs in the dark.) Removal of the nucleus from very young plants provokes premature cap formation, so evidently the information required for morphogenesis is present from an early stage of growth even though it is not expressed in an entire plant until it reaches a certain size (2–3 cm). Such observations suggest that the nucleus is not directly concerned with morphogenesis, just as it was found not to be necessary for biological clock activity. However, Hämmerling found that it was possible to graft stalks and rhizoids of different species, having distinctively different types of cap structures, together so as to generate a situation of conflict between two types of morphogenetic information. If, for example, the stalk of *Acetabularia mediterranea* from which cap and rhizoid have been removed is grafted to a rhizoid of *A. crenulata*, then the cap which regenerates is that of *A. mediterranea*. However, if this cap is now removed and a second one allowed to grow, it begins to take on the morphological characteristics of the grafted nucleus, *A. crenulata*. If this second cap is amputated, the next cap is completely of nuclear type. This experiment is the morphogenetic analogue of the clock experiment in which a stalk 180° out of phase with a rhizoid and

Figure 5.3. A photograph of *Acetabularia mediterranea* showing the mature cap, the stalk and the rhizoid which contains the nucleus.

its nucleus are grafted together, with the result that after several cycles the phase of the nucleus dominates. Such observations tell us that whatever the cytoplasmic mechanisms are for constructing a morphogenetic map (gradient system) or running a clock, they can function independently of, but are responsive to, nuclear influences. It is evidently the cytoplasm and its organelles that are carrying out the organization of events in space or in time.

Extensive studies by Brachet and his colleagues (see Brachet, 1964; Brachet and Bonotto, 1970) on some biochemical aspects of morphogenesis in *Acetabularia* have established that a substance necessary for apical cap formation is sensitive to u.v. irradiation and ribonuclease treatment and that its appearance can be inhibited by actinomycin D. These observations point conclusively to the involvement of RNA in the morphogenetic process, as we would expect. Furthermore, they have also established that the apex is the main centre of cytoplasmic protein synthesis and that polyribosomes are most abundant in this region, although protein synthesis occurs throughout the cytoplasm and goes on in plant sections which fail to regenerate caps. The problem of the spatial organization of epigenetic processes is further complicated by the fact that there is active cytoplasmic streaming or cyclosis which effectively mixes the contents of the cytoplasm throughout the plant from apex to base; yet morphogenetic information is distributed in a stable gradient. This gradient is also stable to centrifugation of the plant, which redistributes the cytoplasm transiently; or to gentle removal of the cytoplasm from a stalk segment and its replacement by the same or other cytoplasm (Sandakhchiev *et al.*, 1972). Evidently the gradient must be located on structures which are not disturbed by such treatment, such as the membranes or the cortex adjacent to the polysaccharide cell wall, and it is at this point that we can turn to the ciliate protozoa for further evidence in relation to the role of the cell surface in morphogenesis.

Some of the relatively large ciliates such as *Stentor*, *Tetrahymena*, *Paramecium*, *Euplotes*, *Blepharisma* and *Spirostomum* have been studied in considerable detail in an attempt to gain some insight into the mechanisms governing regeneration and morphogenesis in acellular systems. Their spatial patterns can be very complex, consisting of oral structures, internal organelles and cortical structures such as cilia, flagellae, trichocysts, etc., all distributed in highly-defined patterns. Small fragments of the organism are capable of reconstituting the whole; while there are particular regions of an organism which exert a dominant organizing influence on the rest of the cell, inducing coherent spatial order, even after considerable surgical deformation. This region has been called the 'organization area' (Jerka-Dziadosz, 1964). By interfering surgically with the orientation and relationships of parts of the ciliate cortex, and by studying the heritable stability of anomalous patterns of organization of cortical structures, it has been possible for workers in this field such as Sonneborn (1963), Tartar (1967) and Nanney (1968) to develop an interesting picture of the significance of the cortex in morphogenesis and its relation to nuclear information. A recent review of the material by Frankel (1974) and a detailed analysis of morphogenetic mechanisms in *Chilodonella cucullulus* by Kaczanowska (1974) present very convincing arguments for the case that the principles of spatial organization in ciliates are essentially identical with those in multicellulars, and that the basic mechanisms operate in the cortex.

This insight is not a contemporary one, having been recognized and explicitly expressed by Driesch as long ago as 1908 and more recently by Stumpf (1968). Frankel shows how the postulate of gradients from oral primordia acting as organizing centres resolves observations on disturbed ciliary and buccal patterns resulting from transplantation and excision experiments and also those relating to nuclear influences on these patterns. Once again what emerges is that the cytoplasm, and more particularly the surface structure associated with the cortex, is the site of spatial organizing activity which is responsive to specific nuclear information but acts independently of it. So we must now see if we can account for such behaviour in terms of simple and general properties of biological surfaces in order to explore the possibility of laying a universal foundation for the primary events in morphogenesis that apply to unicellular as well as to multicellular organisms.

Membranes and Morphogenesis

It is now known that biological membranes are more than phase boundaries between organelles and cytoplasm, or between cytoplasm and external environment. The classical Davson and Danielli (1952) lipoprotein model, with a lipid bilayer coated on either side by protein, still serves as a basis of membrane structure; but this has been much extended by studies such as those of Bretscher (1971) on the existence of proteins which span the membrane and by the demonstration that many enzymes such as adenosine triphosphatase, dehydrogenases and adenyl cyclase are membrane-bound and function most effectively in this state (e.g. Porcellati and di Jeso, 1971; Siekevitz, 1972). Thus there is considerable metabolic activity associated with membranes, in addition to their familiar electrophysiological properties and selective ionic permeability. The phenomenon of electrical excitation is a well-known feature of many biological membranes, but we should see this as only one form of excitability which includes metabolic activity as well. By this I mean that the presence of allosteric enzymes, capable of being activated and inhibited by diffusible ligands, makes possible the occurrence of activity waves on membranes similar to propagating action potentials, but not necessarily accompanied by membrane depolarization and the flow of ions across the membrane. I will now describe a simple model of a metabolically excitable membrane which can serve as a basis for unifying the morphogenetic phenomena considered so far in this chapter and which we can then use as a basis for analysing the more complex morphogenetic systems which will be considered later.

Suppose we have an enzyme, E_1, which is bound to the membrane or cortical layer of a cell with its active sites facing the cell interior, as shown in Figure 5.4. This enzyme catalyses the transformation of a metabolic species

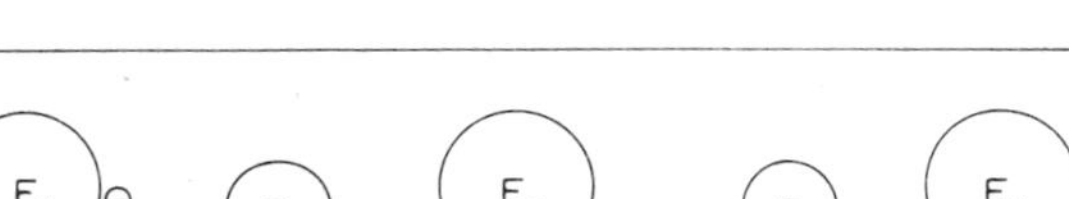

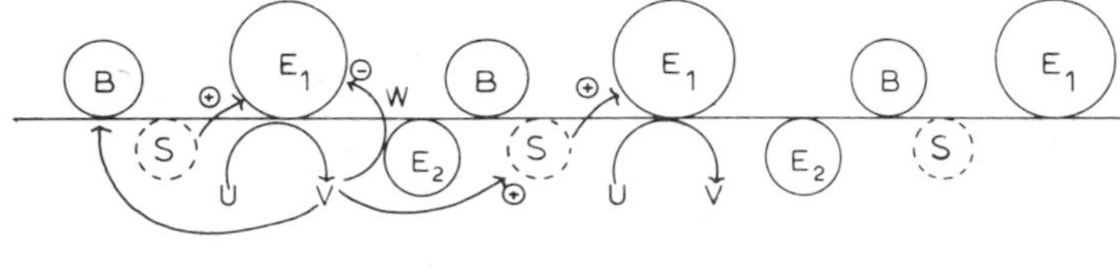

Figure 5.4. The pattern of interactions of molecular processes in a membrane or a cortical layer whose activities can result in the generation of a gradient of the metabolite V bound to the proteins, B.

U into another species V. A second enzyme, E_2, which may or may not be bound to the membrane, catalyses the reaction V to W. Assume that the metabolite V causes an increase in the concentration of an activator of enzyme E_1. This is shown in the Figure as S. This activator could be released from bound sites by V, or in the case of an ion such as K^+ the increase could arise from a change in membrane permeability effected by V, or by some other alteration of ion balance locally. This activation of E_1 results in the existence of a positive feed-back loop: the more V is produced, the more free S occurs and the more will E_1 be activated to produce yet more V. This results in the property of metabolic excitation, since if free S is produced transiently at some point in the membrane, a wave of activation of E_1 can spread along the membrane from the point of initiation. To stabilize the process, assume that W is an inhibitor of E_1. The reader will recognize that the circuitry of the control system is qualitatively similar to that discussed in Chapter 1 in connection with the glycolytic oscillator (Figure 1.10) and that the system described above has the potential of undergoing metabolic oscillations; i.e., the excitation wave could recur spontaneously from an origin. If one of the ligands involved in the control system is an ion, then an electrical signal will accompany the wave. Otherwise it will be an electrically neutral metabolic process.

The next question is how one can get a redistribution of matter as a result of the activity wave so that a stable gradient of bound metabolite is formed on the membrane, leaving a spatially-ordered 'memory' trace of the wave. If we suppose that there are on the membrane binding sites for the metabolite V, designated by B in Figure 5.4, and that these sites are activated by W, then it will follow that after the passage of a wave, bound V will be distributed in some manner on the membrane. To ensure that the morphogen, now identified as substance V bound to sites B in the membrane or cortex, has its maximum at the wave origin and elsewhere decreases monotonically, certain quite simple assumptions can be made. For example, if there is a sufficient delay between the production of V and the activation of B, then a monotonic gradient results, as shown in the Appendix. This delay is achieved simply by assuming a slow conversion of V to W and a threshold level of W for activation of B. Another possibility is for S to act as an inhibitor of B as well as an activator of E_1. Then as the wave propagates, S (assumed to be an ion) accumulates almost uniformly within the system because of its large diffusion constant, and there will be a progressive inhibition of the binding reaction as the wave propagates. The extent of the inhibition at any moment in time, hence for any spatial location of the activity wave, depends upon the volume of the system, since S accumulates at a rate inversely proportional to the volume of the space into which it is diffusing (assuming closed boundaries). This provides a size-sensing mechanism. The properties of such systems are under study to investigate more exactly their regulative properties, about which more will be said below.

In general, one activity wave will not create a stable morphogenetic gradient. This requires the regular recurrence of waves initiated periodically from the origin. This periodic activity of the origin, which now acts as a pacemaker or organizing centre for the system, was in fact anticipated in the description of the basic kinetic circuitry shown in Figure 5.4. It was pointed out that the feed-back structure of the control processes acting on E_1 is such that oscillatory behaviour in the variables can occur, the period being determined primarily by the rate of disappearance of the inhibitor, W, by degradation and diffusion. Since bound metabolite will itself also be released at a finite rate, the general picture that emerges for this extended model is that a gradient on the membrane will be maintained as a result of a balance between recurrent waves propagating from an origin and release of the metabolite from the binding protein in the membrane. Such a system is, of course, a labile or plastic one which can either maintain its own structure if undisturbed, or respond to external influences inducing a change of spatial order.

The wave model as described here belongs within the category of systems for which there exist propagating wave solutions of diffusion-reaction equations such as equation (5.1). This is an area of active research at the moment and it seems likely that experimental work on wave behaviour in developing systems together with the analytical study of model systems capable of exhibiting wave solutions will lead to an extremely interesting new approach to the understanding of embryological processes. The membrane model we have been considering has been treated in an essentially phenomenological manner, the possibility of wave propagation being assumed and its consequences as regards gradient formation explored. There are now in the literature sufficient examples of both experimental and theoretical systems displaying the properties of wave propagation for this to be taken as an established feature of chemical and biochemical systems. A number of examples may be found in a recent Faraday Society Symposium (1974), which constitutes a timely survey of work in this very interesting field of study. Let us now see how the general picture of wave propagation and its association with gradient formation corresponds to observations on certain developmental systems which provide important insights into the earliest events associated with the establishment of polar order.

The eggs of the intertidal algae, *Fucus* and *Pelvetia*, are spherical, about 70 μm in diameter and are fertilized by minute sperm after being shed into the sea. There is no evidence for any polar axis in the eggs at this time. The eggs sink after fertilization and attach to any available substratum (usually rocks) by means of a sticky mucopolysaccharide which they secrete. Fertilization or parthenogenetic activation of these eggs is readily effected in the laboratory. The first sign of asymmetry occurs some 15 hours after sperm entry and consists of an outgrowth at one pole. The egg becomes pear-shaped and then divides at right angles to the long axis, producing a smaller rhizoid cell which includes the outgrowth and a larger thallus cell. The former develops into the holdfast and the latter into the fronds of the plant.

A number of stimuli are capable of determining this polar axis. Lund (1923) showed that electrical fields can do so, while Whitaker (1938) demonstrated the effectiveness of a pH gradient. Other stimuli include a temperature gradient and light. Jaffe (1966, 1968) has shown that the rhizoid pole becomes electrically negative relative to the thallus, so that effectively a current travels through the egg. Jaffe considers this to be the initial polarizing force which generates a gradient of substance within the egg by a process of self-cataphoresis, electrically charged molecules becoming spatially distributed within the egg as a result of the electrical polarity.

The evidence is strong that it is the membrane which is the site of the initial polar order.

The wave model described above can account easily for these observations. In order to explain the electronegativity of the rhizoid, we could assume that the bound metabolite, V, has a depolarizing effect on the membrane, assuming that the wave originated from that part of the egg which gives rise to the rhizoid so that the maximum level of bound V occurs at this point. This is in accordance with the observation that if eggs are placed in a gradient of KCl, the rhizoid develops towards the high point of the gradient, and is consistent with observations on the effects of electrical fields and pH gradients. The effect of light is not explained, although one could assume that light receptors in the membrane affect it in such a way as to cause hyperpolarization, since the rhizoid grows away from the light.

It is of interest to our enquiry that Nucitelli and Jaffe (1974) have observed a spontaneous action potential in developing *Pelvetia* eggs that recurs periodically, originating from the rhizoid. A similar action potential has been reported to occur in regenerating *Acetabularia* segments by Novak and Bentrup (1972), the electrical signal originating at the apex. This is first observed to occur some 12 hours after decapitation, with a somewhat irregular periodicity of about 1 per hour, the period decreasing as regeneration proceeds to about 1 every 10–20 minutes. There is at present no experimental evidence that these electrical waves are in any way connected with the establishment of substance gradients in these developing systems, but we are currently looking into this possibility. The general correspondence between the behaviour of *Pelvetia* and *Acetabularia* and that anticipated from the wave model of gradient formation is quite clear.

Let us now see how we could apply this model, in an appropriately modified form, to the behaviour of myxamoebae of *Dictyostelium discoideum.* Let us suppose that $S = K^+$ and that one of the changes in cell state occurring after amoebae stop feeding is the development of excitability in the membrane, due perhaps to an increase in the concentration of E_1 in the membrane or a change in the intracellular K^+/Na^+ ratio. Assume that cyclic AMP, the chemotactic signal, has the effect of initiating an activity wave by some means such as altering membrane permeability to an ion which affects free intracellular K^+, the result of a cAMP stimulus being an increase in the concentration of free potassium ions within the cell.

Now the activity wave in a myxamoeba must travel faster than the diffusion rate of the chemotactic signal, which has been estimated to be about 40 μm/s (Gerisch, 1968), since otherwise all parts of the amoeba membrane will be activated by the diffusing signal before a wave has had time to propagate and establish a gradient. There would then be no wave origin.* Meinhardt and Gierer (1974) fail to take account of this in discussing the application of their model to *Dictyostelium* amoebae. An activity wave travelling with a velocity greater than 40 μm/s could be of the type considered above, involving the activation of an enzyme in the membrane, but it seems more plausible that such a wave would be purely ionic in nature and of the classical electrochemical variety, as in a neurone. In order to obtain a gradient from such a wave, one could suppose that the binding proteins in the membrane, *B* (Figure 5.4), are capable of picking up sodium, which would then play the role of the diffusible metabolite *V* in the original model. Thus as the wave of depolarization travels along the membrane from the point where a super-threshold concentration of cAMP is first received, a membrane-bound gradient of sodium will be left as the memory trace of the wave. Furthermore, the membrane-bound sodium gradient could itself be responsible for the polarity of the amoebae, pseudopods tending to form preferentially at the partially-depolarized region of the membrane (elevated intracellular sodium causing the depolarization). Alternatively, the membrane-bound sodium ions could have an indirect effect on pseudopod formation by some pattern of interactions within the membrane, resulting in a weakening of forces between structural elements and consequent outflow of cytoplasmic material into pseudopods. In either case one has a polarized amoeba which will tend to move towards the source of cAMP. Furthermore, the mechanism of polarization is a labile one, its duration being determined by the stability of the gradient on the membrane so that repolarization by signals can occur when the amoeba drifts off course.

There is a delay of about 15 seconds between reception by an amoeba of an effective cAMP signal and a pulse-like release of cAMP from the cell. It is not clear whether the controlling process in the generation of the pulse is the rate of intracellular synthesis of cAMP or a release mechanism involving the bursting of vesicles containing the chemical as occurs in the neurotransmitted release process in neurones.

During the later stages of aggregation cell to cell contacts are formed. These are necessary for the formation and development of the slug (Beug *et al.*, 1973) and it is evident from the behaviour of individual amoebae during aggregation that one aspect of their polarity is a differential distribution of adhesive sites over the cell, stable contact being made preferentially between the anterior of one cell and the posterior of the next. Thus another consequence of the gradient postulated to form within single cells must be a polar distribution of membrane proteins acting as adhesive units on the cell surface. Once cells have made contact, waves can propagate continuously from cell to cell by the passage of ionic signals via low resistance or gap

* I am indebted to Steven Ludlow for pointing this out to me.

junctions. A gradient could be formed over the whole organism if V can diffuse across the junctions. Periodic behaviour would continue over the slug, dominated by the tip which acts as an organizer, as described by Rubin and Robertson (1975). A model which utilizes the polar distribution of 'contact-sensing' molecules on the amoeba membranes as the basis for the establishment of a gradient of cAMP over the slug is given by McMahon (1973). A limitation of this model is its failure to show regulation after cutting the slug in half, and its dependence upon rather special initial conditions to give the 2:1 spore:stalk ratio in undisturbed slugs.

Regulation and Stability

It is of interest to ask what the biochemical nature of the molecular species involved in the wave model are likely to be. The most interesting of these is V itself, the molecule which when bound to active sites on membranes plays the role of the elusive morphogen. Since I have assumed that this molecule affects the ion balance and hence is implicated in the control of membrane properties, it is evident that we need to consider those metabolites which can have such an effect. The trouble is, there are very many. However, the requirement of generality and universality for the model restricts our choice somewhat. One interesting candidate for V is cyclic AMP itself, in which case E_1 would be adenyl cyclase, known to be associated largely with membrane fractions, and E_2 would be phosphodiesterase, a soluble enzyme. Instead of affecting membrane permeability directly, cAMP could exert its primary effect by activating protein kinases, thus generating a secondary gradient of covalently-bound phosphate on the membrane. This could result in two important consequences. First, the stability of protein distribution in the membrane could be increased as a result of phosphorylation and consequent electrostatic interaction and hydrogen bonding between membrane constituents. This would reduce the mobility of membrane proteins and so stabilize the gradient against diffusive degradation as envisaged in the fluid mosaic picture of membranes (Frye and Edidin, 1970; Nicolson and Singer, 1971; Gitler, 1972). Second, phosphate groups on membrane-bound proteins would tend to bind divalent cations such as Ca^{2+}, which in turn could have the effect of opening Na^+ or K^+ channels in membranes by a mechanism similar to the proposed action of rhodopsin in the cones of the retina (Hagins, 1972) and troponin in muscle (Bremel and Weber, 1972; Hitchcock, 1973). The metabolic consequences of local changes in ion concentrations mediated by membrane-bound proteins are now recognized to be profound, since this appears to be one of the primary modes of action of hormones.

Cyclic AMP is only one of the candidates for the substance V of the model, others including cyclic GMP, serotonin (5-hydroxytryptamine), γ-aminobutyric acid, etc. These are all simple metabolites which could play important primary morphogenetic roles as membrane modulators and hence affect local metabolism in specific ways which result in ordered physiological and structural heterogeneity within cells or developing tissues. They could be part of the universal language of pattern formation in organisms.

The feature of the present model which suggests that it may have regulative properties is that the processes involved in gradient formation are essentially local ones; i.e. the metabolite V diffuses from its site of production on membrane-bound enzymes E_1 and is picked up by proteins B lying nearby in the membrane. Once the wave-front is a certain distance from any particular region on the membrane, very little diffusing metabolite will be picked up because the binding sites will have relaxed back to an inactive state. Thus we may anticipate that the level of morphogen at the origin will be close to its maximum value once the wave has propagated a certain distance. This will, of course, depend upon the values of v and D, smaller D values producing more local behaviour.

An important factor in this process is the nature of the pick-up function itself, describing how long it takes for binding sites to be activated and then to relax to an inactive condition. Using a function of the form

$$f(z) = z\,e^{-\alpha z}, \qquad z \geqslant 0$$
$$= 0, \qquad z < 0$$

with α such that the binding process reaches its maximum 20 μm behind the point where activation of B begins and then falls off exponentially, it has been shown (see Appendix) that with $\nu = 5\ \mu s^{-1}$ and a delay of 4·5 s between wave initiation and pick-up, the morphogen reaches 96% of its maximum value at the origin of the wave when this has propagated about 60 μm. Assuming that 96% of the normal range of morphogen values is functionally equivalent to the complete range, using a purely arbitrary figure, one can then conclude that in the wave model, for the above parameter values, all axial lengths greater than 60 μm will have the same effective range of morphogen values. Lengths less than this will be deficient, i.e. regulation will fail. Different organisms will, of course, have different parameter values, so that the length at which size independence fails will differ among them. We see that the wave model gives us regulation as a bonus: no special assumptions need to be added to the basic ones to give us this biologically important property. Thus the wave propagation process has the advantage over the class of models described by equation (5.1) that regulation follows as a natural consequence of the basic postulates. Furthermore, the membrane wave model explains how morphogenetic field information can be stable to disturbances such as centrifugation of the cytoplasm; removal, homogenization, and replacement of the cytoplasm; and how the field can be affected by rotation or transplantation of segments of the cortical layer. Such observations are not easily explained by models of enzyme activity and metabolite diffusion in the bulk phase of the type described by Turing (1952) or by Gierer and Meinhardt (1972). Let us now turn to the question of the experimental implications of the wave model in relation to hydroid morphogenesis.

Periodic Behaviour in Hydroids

There are two aspects of the general theory which is being developed here for morphogenetic processes which lend themselves to experimental investigation of a fairly distinctive type. One is the expectation that membranes are primary sites of early morphogenetic processes, so that the properties of biological surfaces such as their electrical excitability, their responsiveness to a variety of membrane-active agents and changes in membrane composition such as the covalent binding of phosphate groups to membrane proteins would provide experimental approaches to early morphogenesis. The other is the expectation that recurrent periodicities, possibly associated with ion flows and hence electrical signals, would be a significant aspect of formative processes in developing systems. So far we have seen that in *Dictyostelium discoideum* there are clear periodicities associated with aggregation, and in *Pelvetia* and *Acetabularia* there are recurrent action potentials correlated with the early stages of morphogenesis. Is there evidence of periodicities in hydroid morphogenesis?

Time-lapse films of regenerating *Tubularia* taken by J. Cooke and G. C. Webster at the Marine Biological Station in Naples showed the following sequence of events. After cutting off the hydranth, there is first a migration of cells distally and a tendency for them to condense in a region slightly proximal to the cut tip. This phase lasts for about 5 hours. Then a phase of irregularly periodic contractions begins with contraction waves spreading out first from different centres in the more condensed zone and finally assuming a coherent, organized, periodic propagation from a single pacemaker region with a period of eight to ten minutes. This lasts for some eight and a half hours and is followed rather abruptly by a quiescent phase of about four and a half hours during which there is a further condensation of cells in the presumptive tentacle region and the beginnings of morphogenesis. The pacemaker region now coincides with the centre of this condensed zone. Another phase of periodic contractions and wave propagation occurs, lasting about five hours, followed by another quiet period. Unfortunately the film sequence was not complete from this point up to full differentiation of the hydranth, but the last phase of regeneration consisted of more rapid contractions (four to five minute periods) with a final pulsatile emergence of the completed hydranth from the perisarc with active tentacle movements after emergence. Rather similar observations have been reported by Beloussov *et al.* (1972) on *Obelia*. It is unfortunate that such studies cannot be made on regeneration in *Hydra* because vigrous contractions of the body recur every few minutes and mask all morphogenetic movements. It is the *Tubularia*-type of periodic behaviour which one anticipates as an accompaniment to morphogenesis on the basis of the wave propagation model, and for which aggregation in *Dictyostelium discoideum* is the prototype. However, it is important to realize that metabolic periodicities need not result in observable periodicities of cell movements in developing systems and it may in fact be very difficult to detect the proposed periodicity of metabolic activity on the membrane.

An alternative approach to the experimental study of periodicities in developing systems is to perturb the system by some periodic signal. This procedure was used in *Hydra littoralis* in the following manner. The hypostome was cut off large, well-fed animals and then a second cut was made distal to the budding zone (to avoid complications arising from buds). These isolated gastric sections were then impaled on a 40 μm platinum wire which was insulated except for about 4 mm at the end. The wire was passed diametrically through the organism (now a hollow cylinder), positioned either distally, near the cut that removed the hypostome, or proximally, near the region where the budding zone had been. Two to four animals could be impaled on the same wire, which served as one electrode for a stimulation circuit. The other electrode was another piece of platinum wire which was placed about 1 cm from the impaled animals, everything being immersed in *Hydra* medium whose composition is approximately that of fresh lake or stream water. The wires were connected to the terminals of a Grass stimulator which delivered a DC pulse whose voltage and duration could be selected. It was found that isolated gastric sections contracted in response to a brief (100 ms) four to five volt pulse, but weaker pulses did not produce visible contractions. Since contractions were not desired, the voltage was set at 1·5 V, which was found to cause a significant disturbance of morphogenesis when the frequency was within a particular range. Frequency was controlled by a Crouzet interval timer which activated the stimulator at set intervals. The duration of the pulse was set at 150 ms, since it was found that pulses less than 100 ms were less effective in eliciting morphological perturbations, whereas pulses greater than 100 ms were all equal in their effects, up to 3 s, which was the longest pulse tested. The pulsing was started within one hour of impaling the gastric sections and lasted for 24 hours. At the end of this time, animals were removed from the wire and transferred to separate dishes. The experiments were run at temperatures between 21 and 24 °C.

Three types of controls were run: (1) mid-gastric sections were allowed to regenerate undisturbed: (2) sections which were impaled for 24 hours but not stimulated: (3) sections stimulated at 1·5 V for a period of time equal to the total duration of stimulation during the 24 hours, then allowed to regenerate undisturbed. Thus if the frequency was one pulse every two minutes, the continuous stimulus used in control (3) was:

$$\left(\frac{24\times 60}{2}\times 0{\cdot}15\right)=108\text{ s.}$$

Impaling animals for 24 hours had no effect on regeneration except to produce a delay of 12–16 hours (measured to appearance of tentacles), which was within the normal range of variability. Animals stimulated continuously for 108 s or longer (maximum three minutes) showed no response.

The morphogenetic disturbance which resulted from stimulation proximally at frequencies in the vicinity of a half minute is illustrated in Figure 5.5D. The animal has a double peduncle and a medial hypostome. The pictures in the series show the same animal (photographs not all at same magnification) on days 2, 3, 4 and 5 after the end of the 24-hour stimulation period. When a similar treatment was applied distally, the frequency of occurrence of this abnormality was very much smaller. The spontaneous frequency observed with the controls was 1·5%. These results, showing the responses to stimulation at different frequencies are given in Table 5.1. Not all organisms scored as significantly abnormal were identical in appearance with that in Figure 5.5D, but they all had two peduncles and a single medial hypostome. The frequency response curve of the proximally-stimulated animals is shown in Figure 5.6. Maximal response occurred at a frequency of 1 pulse every 2·5 minutes.

An interpretation of these results in terms of the wave model is obvious: the periodic stimulus initiates an activity wave and so induces a gradient along the axis of the organism. The fact that distal stimulation produces a low incidence of abnormal animals while proximal stimulation causes a much larger number is due to the fact that distal stimulation reinforces the normal hypostomal-peduncle gradient while proximal stimulation opposes it. The results obtained are very similar to those observed by Wilby and Webster (1970) in consequence of hypostome grafts to proximal ends of

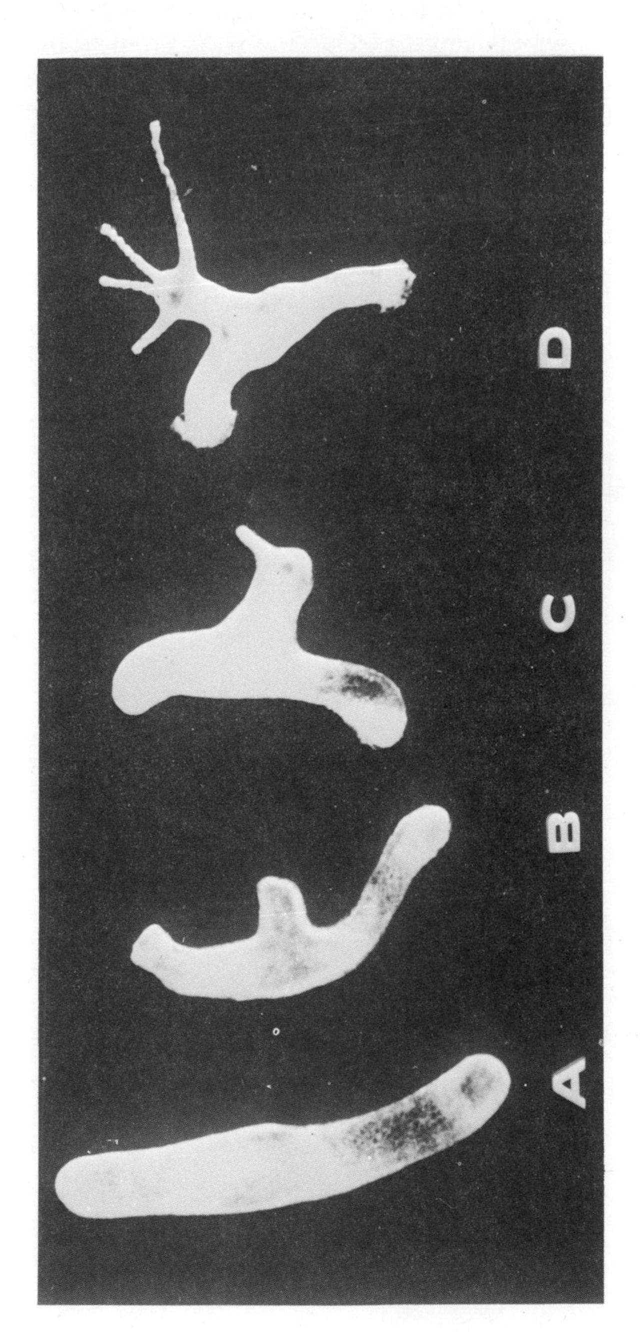

Figure 5.5. Appearance of a medial hypostome in a proximally-stimulated *Hydra*, showing growth on successive days from 48 hours after stimulation.

TABLE 5.1

	Proximal stimulation			Distal stimulation		
Frequency	Total no. of animals treated	No. with double peduncle	Per cent	Total no. of animals treated	No. with double peduncle	Per cent
$\frac{1}{5}$ min	19	1	5·3	14	0	0
$\frac{1}{3}$ min	18	0	0	20	0	0
$\frac{1}{2\cdot5}$ min	42	13	32	38	2	5
$\frac{1}{2}$ min	22	5	23	36	1	3
$\frac{1}{1\cdot75}$ min	22	2	9·1	22	0	0
$\frac{1}{1\cdot5}$ min	18	2	11·1	16	0	0
$\frac{1}{1\cdot25}$ min	11	0	0	10	0	0

Frequency of spontaneous double peduncles and medial hypostomes observed in controls: 3 out of 200 = 1·5%.

isolated digestive zones. Such grafts left for 48–72 hours and then removed also resulted in animals with double peduncles and a medial hypostome. Proximal grafts left in place longer than 72 hours before removal caused a

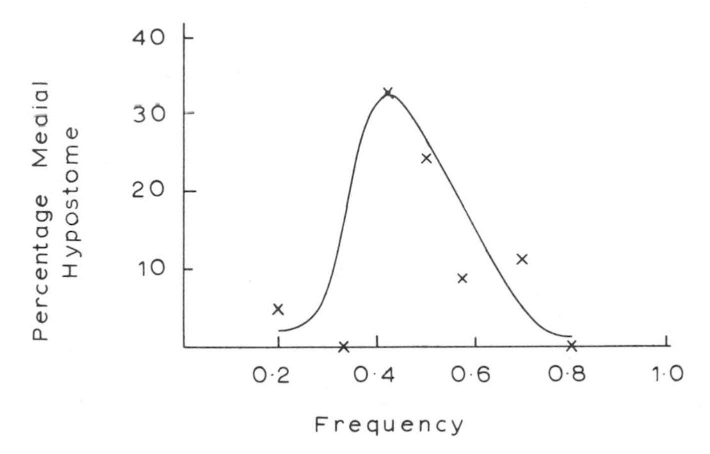

Figure 5.6. Frequency response curve for medial hypostome generation after proximal stimulation.

more complete polarity reversal in the animals. In the electrical stimulation experiments, animals kept on the electrode and stimulated longer than 36 hours or so showed such extensive general damage that their regeneration was very abnormal, or did not occur at all. This indicates that the electrical stimulation was a very crude and relatively destructive stimulus, and that as a simulator of a pacemaker it is poor. However, the fairly sharp frequency response at 1/2·5 min and the relative effectiveness of proximal in contrast to distal stimulation suggest that the period and the position of the stimulus are specific, indicating that some kind of pacemaker is operative in *Hydra*. This type of experimental design, with refinements, could provide insight into the position and nature of pacemaker centres in developing organisms. A beautiful example of the use of a chemical periodicity to control morphogenesis is provided by the experiment of Robertson, Drage and Cohen (1972) in which periodic pulses of a cAMP were delivered to a lawn of amoebae, which then aggregated to this artifical pacemaker.

Evidently a periodic model of the type we are considering can be used to explain morphogenetic processes which are either overtly periodic in time or are continuous, since the underlying periodicity may not be easily observed. Thus it is not very profitable to go further with a description of developing systems in which temporal periodicities of one kind or another have been observed in order to present evidence in its favour. The utility of a theory is not so much its consistency with current evidence, since such consistency can nearly always be demonstrated by one means or another, but rather its intellectual appeal, its simplicity and hence its capacity to provide insight into basic principles operating behind a diversity of processes whose particular characteristics obscure the general pattern. In the end, of course, it is the organism that is the final arbiter, since they will employ developmental strategies which suit their particular needs rather than our requirements of intelligibility.

Chapter 6

MORPHOGENESIS: SPATIAL PERIODICITIES AND MULTIDIMENSIONAL ORDER

SPATIAL periodicity is a prominent feature of biological pattern and morphology, and it was in explanation of this that one of the earliest analytical models of morphogenetic processes was advanced. This is A. M. Turing's (1952) analysis of instability in coupled diffusion-reaction processes, already referred to in the last chapter in relation to the equations proposed by Gierer and Meinhardt to explain polarity in *Hydra*, which belong within the same general category. The type of biological phenomenon which Turing sought to explain is the pattern of spots on a Dalmation, the periodic segmentation of annelids, or the circle of tentacles in hydroids. The spotting pattern in a Dalmation is a two-dimensional problem while the others involve periodicity along a line or a circle.

Assuming that the biochemical processes underlying pattern formation involve the interaction of two molecular species, present in concentrations X and Y, and that these can diffuse throughout some spatial region, the general form of the equations expressing the behaviour of X and Y as functions of space and time are:

$$\left.\begin{aligned}\frac{\partial X}{\partial t}&=f(X,Y)+D_X\nabla^2 X\\ \frac{\partial Y}{\partial t}&=g(X,Y)+D_Y\nabla^2 Y\end{aligned}\right\}. \qquad (6.1)$$

Here D_X and D_Y are the diffusion constants for X and Y, whilst ∇^2 is the differential operator $(\partial/\partial x^2+\partial/\partial y^2+\partial/\partial z^2)$ describing the diffusion process, x, y and z being spatial co-ordinates. Turing showed that there are functions $f(X, Y)$ and $g(X, Y)$ all of whose steady states are unstable. He did this by linearizing the functions in the neighbourhood of the steady state and by assuming that the system was partitioned into cells which were arranged in a circle or on the surface of a sphere. The instability of the steady state means that small fluctuations in the system will cause it to diverge more and more from the homogeneous condition of spatial uniformity. Turing showed that the system then developed a wave-like distribution of the chemicals X and Y, but he did not show that the system would necessarily evolve to a stable non-homogeneous steady state with a defined spatial periodicity. This is because equations such as (6.1) with non-linear f and g cannot be solved analytically beyond the neighbourhood of the homogeneous steady state and computation is necessary. Although Turing himself was very much involved in the early development of computers and has given his name to the logical embodiment of recursive processes, viz, the universal Turing machine, computational facilities were not nearly so available then as they are now.

Complete solutions of the equations (6.1) using explicit forms of the reaction functions f and g have only recently been computed, but they fully confirm Turing's conclusions. Martinez (1972), for example, used the functions

$$f(X, Y) = a + X^2 Y - (b+1)X$$
$$g(X, Y) = bX - X^2 Y,$$

which were originally suggested by Prigogine (1969), and showed that for a linear array of cells stable spatially periodic solutions of X and Y resulted. Turing had demonstrated that in general the wave-length of a spatial period would be constant for given parameters, so that the larger the linear dimensions of the system, the more spatial cycles there would be. This was also confirmed by Martinez, who showed further that the amplitude of the waves varied as a function of overall linear dimensions. Another recent computer study of this type of system was made by Bard and Lauder (1974), confirming the above observations and demonstrating also that the patterns obtained are not reliable: small variations in initial conditions give significant changes in the number and positions of the peaks. Thus systems with spatial heterogeneities arising from the Turing mechanism are neither reliable in generating spatial order, nor do they regulate. For many morphogenetic processes, such as tentacle formation in hydroids, this is a perfectly satisfactory model, since the number of tentacles increases with the diameter of the body. However, in periodic systems which show regulative capacity, such as the digits of vertebrate limbs, the model needs modification to give constancy of period number and of amplitudes in order to achieve reliability and invariance of positional information. An interesting analysis of this problem is given by Maynard Smith (1960). Regulation requires length-dependent changes in the parameters of the model, which presupposes some kind of auxiliary size-sensing mechanism.

Wave Propagation and Spatial Periodicities

Let us now consider an extension of the wave propagation model of Chapter 5 which gives spatial periodicities. One way of approaching this problem is to suppose that at regularly-spaced points along one or more dimensions of a tissue (depending upon the nature of the periodic pattern being considered), sources or centres of morphogen production are generated according to some well-defined rules so that there is a resulting periodic distribution of morphogen. In terms of the wave propagation model, this would involve the establishment of wave origins at regularly-spaced intervals along an axis (considering now the case of a one dimensional periodicity). Perhaps the simplest way of achieving this is as follows.

Suppose that there is a wave of competence with respect to the property of excitability which travels over a developing tissue, such that only after the passage of the wave does the tissue become capable of transmitting an activity wave of the type discussed in the previous chapter, illustrated by the reaction sequence of Figure 5.4. It is therefore assumed that the developing tissue already has an axis which determines the direction of the movement of this wave. The development of this kind of excitability is treated analytically in the next chapter as a phase transition, and the conditions governing this transition are described there. Essentially, this is controlled by the synthesis of the enzyme E_1 and its entry into the membrane, so we are considering a type of epigenetic induction wave travelling over a tissue.

Let us next assume that when it is above a particular threshold value, morphogen inhibits wave propagation, for example by interfering with the activity of enzyme E_1. This means that we are considering the form of the model in which a single activity wave is sufficient to set up an adequate gradient, like that shown in the Appendix. This is because once the wave has occurred and the morphogen gradient is generated, all parts of the membrane carrying super-threshold values will be inhibited from further wave-propagation activity. We are also assuming that the morphogen (i.e., the BV complex) is very stable, taking a long time to decay.

These assumptions result in the generation of a periodic spatial distribution of morphogen. The wave of excitability will travel over the tissue and successive origins of morphogen-producing activity waves will occur at regular spatial intervals, leaving their gradients on the membrane or the cortex. The exact spatial wave-form of the morphogen gradients will depend upon the details of the model, but in general there will be an asymmetric distribution about the peaks, since activity waves cannot propagate 'backwards' into the area with morphogen. Hence the gradient will tend to be steep behind and shallow in front of the new wave origins, the activity wave propagating ahead into tissue left excitable by the induction process. Thus the tissue will tend to have a distinct polarity to its organization, superimposed upon the spatial periodicity of peak morphogen values, as one tends to have in insect segments or in segmenting vertebrate somites.

To get regulation in such a model, as one does in somitogenesis (Cooke, 1975*a*), it is necessary to adjust the inter-peak morphogen values according to the total size of the tissue. In general, this requires a size-sensing mechanism of some kind which adjusts a parameter, such as the morphogen inhibition of E_1, postulated above, in a manner dependent upon overall tissue dimensions. The larger the tissue, the larger the inhibition must be so that the distance between origins of morphogen-producing waves will be greater. There are again various ways of achieving this, one being to suppose that some stable, small and rapidly-diffusing component of the process which originally established an axis over the tissue serves as the regulator. In wave-model terms, this could be a substance such as W, associated with a primary axis-generating wave (hence distinct from the wave which generates the periodically-distributed morphogen). The concentration of this substance present in the cells of a tissue will vary inversely as the total field volume, assuming that it can diffuse from cell to cell within the field, but not across the field boundaries. If this substance antagonizes the inhibition of E_1 by the morphogen, then the smaller the field size the higher will its concentration be and hence the smaller will be the domain of inhibition of a morphogen-generating wave origin, hence the closer the morphogen peaks. This type of model is essentially similar to models of the type proposed by Wigglesworth (1959) for insect bristle distribution and by Thornley (1975*a*,*b*) for the distribution of leaf buds in the phyllotactic process. The additional feature of the present treatment is the use of propagating waves as morphogen sources rather than stationary centres.

A simple and elegant way of resolving the difficulty of achieving regulation in a spatial periodic-generating process has been suggested by Zeeman (1974). It depends upon the initial existence of a regulating monotonic gradient over the system to be partitioned into repeating units and the occurrence of a global oscillation of some variable in the system. Both of these conditions are readily achieved by mechanisms described in previous chapters. The basic postulate of Zeeman's model is the very simple one that the rate of change of cells proceeding along a pathway of differentiation is a function of the mean value of the gradient within a cell, and of the oscillating variable. Thus, for example, cells at positions higher up the gradient may be assumed to differentiate faster than cells at lower values; while during a certain phase of the oscillation this rate of change is speeded up and at another phase, slowed down. Let us suppose further that there is a threshold point in the differentiation process determined by a critical concentration of some metabolite which initiates a switch-like response in cell state. They could arise from the co-operative behaviour of a multimeric protein as discussed in Chapter 1, or from a more complex pattern of co-operative interactions of the type to be considered in Chapter 7. Cells which have passed the switch point may be assumed to undergo a pattern of differentiation different from those which have not. We have then the ingredients of a spatial period-generating process which operates as follows.

The cells nearest to the maximum of the gradient proceed at greatest rate along the pathway of differentiation common to all the cells in the tissue, and so will reach the threshold point first. The global oscillation will at some point go into the phase which slows down movement towards the threshold, and for simplicity we may assume that the process is in fact brought to a halt. However, it need not affect what happens to cells after the threshold has been passed. The result is that some group of cells nearest the high point of the gradient will have passed the threshold point while all other cells are brought to a standstill. The cells past the threshold can then proceed in accordance with their new state, which might be change of cell shape and rotation as in somite formation, or the consolidation of some kind of adhesive and communicational unity which might underlie segmentation in insects. When the global oscillation enters the phase which allows cells to proceed towards the threshold point again, the cells next to the ones which have passed threshold will proceed most rapidly and another block of tissue will have passed threshold when the oscillation once again brings movement towards threshold to a standstill. The group of cells will then undergo the same differentiation process as the first group, but could remain spatially distinct from the latter by virtue of the time difference in entering the new state. A wave of segmentation will thus proceed down the system, cutting off blocks of cells which undergo the same process of differentiation but at different times. This is precisely what happens during somitogenesis in vertebrates. However, it could be applied to a great diversity of spatially periodic phenomena, providing that there is a wave of differentiation of some kind which accompanies the process. Such a wave is a distinctive feature of such a model which distinguishes it from a standing wave of Turing type, which is not initiated from an origin, and from a propagating wave which generates a periodic, membrane-bound gradient unless the wave in the latter process is very slow. Since the wave of differentiation in Zeeman's model is related to the postulated initial gradient, disturbance of the gradient will alter the wave and the spatial periods generated by it and the oscillator. This observation can be used for the design of specific experiments to test the model.

The regulative property of Zeeman's model follows if the initial gradient across the tissue regulates; i.e. if the maximum and the minimum values of the gradient remain the same irrespective of the length of the tissue. Since the successive groups of cells which pass the threshold point do so according to their rate of differentiation as determined by their value in the gradient, maintaining the range of this gradient constant will ensure the generation of the same number of periodically repeating units. However, the size of each unit will vary directly with the overall size of the tissue. Furthermore, providing that the rate of differentiation and the frequency of the global oscillation vary in the same way with respect to rate-determining influences such as temperature, the spatial periodicity will be unaffected even if the overall rate is altered.

Recent studies of somitogenesis in amphibian embryos by Cooke (1975*a*,*b*) and by Elsdale *et al.* (1976) have substantiated the general properties of this model for the generation of spatial periodicities. Cooke has demonstrated the regulative properties of somite formation in *Xenopus laevis* embryos and has provided suggestive evidence for the involvement of a clock-like oscillator by the use of heavy water, ^2H_2O, which causes a significantly smaller number of somites to be formed as compared with controls. Heavy water is known to slow down biological clocks, but its mode of action in somitogenesis need not necessarily follow from such an effect. Elsdale *et al.* have shown that the primary wave of change of cell state which precedes morphogenesis can be perturbed locally by transient temperature changes, a few somites being histologically disturbed by the heat shock, the wave of somitogenesis then returning to the same phase as controls, so that subsequent somites are normally situated, as well as normal histologically. These results indicate that the postulates of regulated global control of the rate of movement of a propagating wave with a temporal periodicity are perfectly plausible, and produce an extremely interesting new approach to the study of spatial periodicities in developing systems as elaborated by Cooke (1975).

The Insect Epidermis

Let us now turn from the relatively abstract problems we have been considering to what biologists tend to regard as real problems, involving some rather more rigorously defined experimental constraints than those which have guided our thinking in relation to spatial periodicities. The insect epidermis has for some years now provided interesting material for the study of morphogenetic fields because it has indicators of tissue polarity whose orientation can be accurately measured under normal and perturbed conditions of development. Very interesting bodies of data are available on disturbances of epicuticular ripples induced by auto- and homograft rotation experiments in *Rhodnius* (Locke, 1959, 1966), on scale orientation in *Galleria* (Piepho, 1955), on bristle and hair orientation in *Oncopeltus* (Lawrence, 1966, 1970), on the spatial pattern of cell differentiation in *Galleria* (Marcus, 1962; Stumpf, 1968) and on the stability of cuticle fibre orientation in the beetle, *Tenebrio molitor* (Caveney, 1973). Such data has sufficient precision of observation to subject gradient models to quantitative test.

A fine example of this is to be found in a paper by Lawrence, Crick and Munro (1972) in which two different models are tested against the observed relaxation patterns obtained by Locke (1959). The experimental procedure was to cut out a small square of integument from a defined position in a particular segment of a larva or pupa, rotate it through 90° and replace it. The rotated piece of integument heals and undergoes development with the host and in the adult the epidermis is examined for disturbances of polarity within the rotated tissue and in its neighbourhood. For the purposes of these experiments, the integument is treated as a tissue with one-dimensional order, since what is observed is simply disturbances of antero-posterior polarity as revealed by any polar marker carried by the tissue. In the case of *Rhodnius* the markers are the ripple patterns in the adult epicuticle, which normally run medio-laterally across the tissue. A 90° graft rotation introduces a conflict between the polarity of the graft and that of the surrounding tissue, resulting in a disturbed ripple pattern within the graft and in its neighbourhood. Complete relaxation would mean that the patterns run parallel to those in the surrounding tissue, while no relaxation would mean that the patterns run at 90° to each other. Experimentally one observes intermediate values, the extent of relaxation depending upon the developmental time at which the graft rotation is carried out, and hence upon the duration of the period during which relaxation can occur. Although the following discussion will be conducted as if the problem were one-dimensional, we will see later that in reality it involves two. And it should also be borne in mind that the insect epidermis is a periodic structure, there being several segments from anterior to posterior, not all identical, but clearly arising developmentally by some periodic space-measuring process. However, we now concentrate upon the problem of field information and polarity within a single segment.

The models used by Lawrence *et al.* (1972) were based upon spatially distributed sources and sinks which maintained a gradient of a diffusible metabolite (morphogen) across the tissue. The first model located the source at the anterior boundary of the segment and the sink at the posterior boundary, giving a linear gradient of the morphogen across the tissue. Graft rotation and replacement would then result in local disturbance of the gradient which would relax at a rate determined by the rate of flow of the morphogen across the tissue, i.e., determined by the diffusion constant and the rate constants for source and sink. This model was computed for particular parameter values and initial conditions and contour plots of the gradient as a function of time were compared with the experimental data by measuring the shape of the contours in the centre of the graft, thus providing an accurate measure of the extent of relaxation. It was found that this model did not give a good fit to the data, the peaks and troughs of the disturbed pattern showing more relaxation than was observed experimentally. This indicated that the graft was more resistant to relaxation than expected on the basis of a purely diffusional process of the type originally envisaged. This was an interesting and valuable result.

There are several ways in which one could alter the system in order to make it 'stiffer', to resist diffusional decay more effectively. The one selected by Lawrence *et al.* (1972) was to suppose that each cell in the tissue acts as a homeostat, attempting to maintain within it a particular level of morphogen. How this level was to be assigned was not specified, but it was assumed that during the developmental process leading up to integument formation cells become 'programmed', presumably by means of a prior positional information field, to hold the morphogen at a particular level, according to their position in the field. Each cell then becomes a source and a sink. These distributed sources and sinks, together with diffusion, would create the gradient of morphogen across the tissue. The 'stiffness' of the system in resisting diffusional relaxation is then determined by the effectiveness of the homeostat in holding its level against diffusional forces causing an increase or a decrease in the level. With particular parameter values, this model was found to fit the experimental data very well.

An additional postulate was introduced in order to account for a further observation. This was that if an insect was induced to moult an additional time beyond the normal, then the disturbed pattern in a graft relaxed even further. There seemed to be a definite correlation between number of moults from the time of the graft operation to observation and degree of relaxation of the pattern. Since a moult cycle involves cell division, it seemed plausible to suppose that the relaxation of the ripple pattern was correlated with some aspect of the cell cycle; e.g. cells might be able to reset their homeostats by 'reading' field information (the level of morphogen in their immediate neighbourhood) at some stage of the cell cycle. This is reminiscent of the quantal mitosis postulate discussed in Chapter 3. Upon adding this property to their model, Lawrence *et al.* obtained a good fit to the relaxation patterns observed after successive moults.

This computer-simulation analysis of a specific gradient model for field information in a developing tissue was of considerable value in several respects, particularly in drawing attention to the necessity for an explanation of the stiffness of the tissue, its ability to maintain a gradient after rotation; and the role of the cell cycle in relation to relaxation. These features have been further explored experimentally by Caveney (1973) in his work on *Tenebrio molitor*. He has presented strong evidence that it is cell division itself which is correlated with the relaxation process. The rotated grafts in the insect also showed a marked difference of relaxation pattern in the anterior as compared with the posterior region, so that some aspect of the process is non-linear, a conclusion already reached by Locke (1959) as a result of observations on 180° rotational grafts in *Rhodnius*, which also show antero-posterior asymmetry. This requires the introduction of further postulates about independent setting of source and sink activities in the Lawrence, Crick and Munro model.

The assumption that the gradient across the tissue is maintained by cells acting as homeostats requires that they be able to set internal metabolite-regulation devices at any one of several different values (say 5). In our discussion of cellular feed-back regulation processes in earlier chapters we did not come across any such multistable behaviour, and indeed it is not easy to design with molecular circuitry, even assuming complex allosteric behaviour. Multistability (i.e. stability of several different states) is undoubtedly a property of developing systems, but I know of no experimental evidence that this exists with respect to a single molecular species within single cells. To require that source and sink activities can be set independently of one another requires yet another order of complexity in the model. Furthermore, Caveney's (1973) observations on cell division point again to the membrane as an important organelle in local polarity determination and in the global relaxation process. One could of course involve membranes in cellular homeostatic activity and polar measurement, but relatively *ad hoc* assumptions would then be required. So it is interesting to see if the principles already introduced in relation to the wave model can be used to explain the observations on the insect epidermis, and perhaps extend the argument somewhat.

The Membrane and the Insect Epidermis

In the description of the wave model in the last chapter it was assumed that the binding sites for V on the membrane were simple, there being one site per protein. However, we could easily assume that the binding protein is a multimer with n sites, and that the kinetics of pick-up is co-operative. In this case the relationship between X, the concentration of V, and the amount of bound metabolite, Y, would be of the type shown in Figure 6.1, which is like Figure 1.8. Now the gradient generated by the wave model is in general non-linear and if the pick-up function has characteristics determined by co-operative kinetics then the gradient will certainly be quite non-linear. In general its shape will be of the type shown in Figure 6.2. We could suppose then, that a segment of the developing integument has a gradient of morphogen of this general shape at the time when graft rotation is performed. Let us assume that wave activity has ceased at this stage, having been responsible for the generation of gradients over the segments by the periodic modification of the wave model discussed above. Graft rotation will then be accompanied by disturbances to the cell membranes during cutting and healing, resulting in the conversion of some V from bound to free form, so that X is produced. This will diffuse locally and some of it will be packed

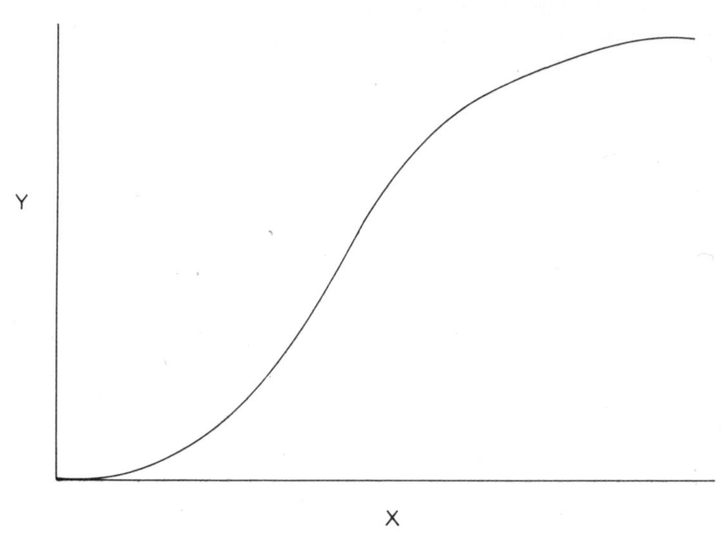

Figure 6.1. The relationship between concentration of diffusible metabolite, X, and bound morphogen, Y, in the wave model if the binding protein is a multimer with co-operative kinetics.

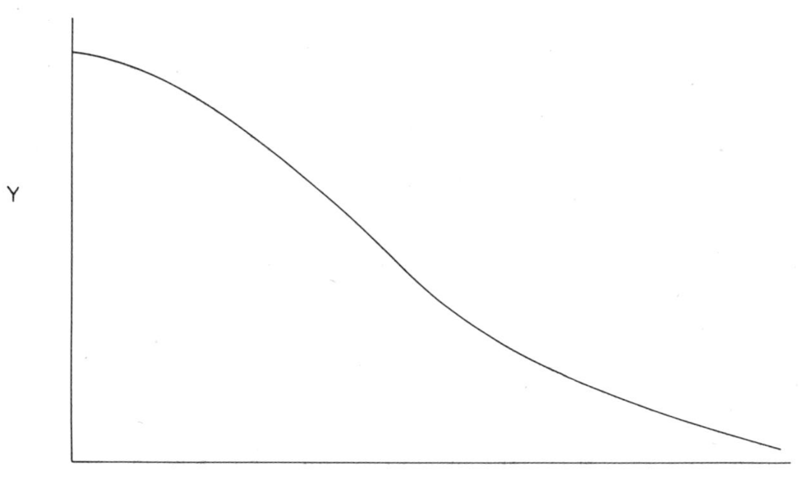

Figure 6.2 The general form of the gradient of morphogen, Y, as a function of distance x, resulting from the wave model.

up again by unoccupied sites on the membrane-bound protein. During the moult cycle, membrane proteins will again be disturbed during the division process and there will be a release of some X and pick-up by unoccupied sites, resulting in relaxation.

The 'stiffness' of the model corresponding to the homeostatic property of the cells in the model of Lawrence *et al* (1972) is the non-linear co-operative pick-up process of multimeric proteins in the membrane, which acts as a kind of store that resists the diffusional relaxation process. At the same time the non-linearities of this model account for the antero-posterior asymmetry of relaxation observed by Locke (1959) and by Caveney (1973), and for the same reason. A detailed study of the correspondences between this model and experimental results is currently under investigation by computer simulation studies.

Most of the ideas about field information discussed so far in relation to the insect epidermis will require fresh scrutiny as a result of recent experimental observations by Bohn (1974), by Lawrence (1974) and by Nübler-Jung (1974). They have shown that the relaxation of perturbed patterns resulting from graft rotation experiments can involve not simply a respecification of field information in cells, but an actual movement of the epidermal cells to new positions within the host field. Furthermore, these ordered movements occur in two dimensions, so that the cells may have two axes of information along which they move to find their 'correct' positions relative to their neighbours. Up to now the problem has been cast in one dimension only, since the antero-posterior direction is the major polar axis in the integument, although there is also clearly a medio-lateral axis of organization as well. However, these recent results require an extension of the whole argument presented above to include cell movement as an aspect of the relaxation process. How this may occur will be considered later when we discuss the retino-tectal mapping in the visual system, where cell fibre growth and movement are primary features of the developmental process. But before involving ourselves in this problem, let us consider a morphogenetic process which appears to proceed by rather different mechanisms from those considered so far, which have all been essentially morphallactic: i.e. morphogenesis involving little or no growth of the system undergoing development. Where growth is an essential feature of the process, the term epimorphosis has been used to describe the morphogenetic transformation. Let us now consider an example of this.

Epimorphosis: The Chick Wing

The development of the chick wing involves the outgrowth, from the embryonic flank, of a tongue of tissue into a paddle-shaped limb bud which extends progressively and undergoes gradual transformation into the characteristic articulated structure of the wing. Saunders (1948) showed that the skeletal components differentiate in proximo-distal sequence and he also established the importance of the apical ectodermal ridge in this temporal morphogenetic sequence. There is some kind of interaction between this most distal strip of ectoderm and the underlying, undifferentiated mesenchyme (Zwilling, 1961) as measured by mitotic index and [^{3}H]thymidine incorporation (Summerbell and Lewis, 1975), until about stage 28 when the limb elements have all been laid down. This actively dividing mesenchymal region, about 300 μm in extent, has been called the progress zone by Summerbell, Lewis and Wolpert (1973), who have constructed a simple model to account for their detailed observations on the consequences of removing the apical ridge and the progress zone at various times of limb development and transplants between chick wings at different stages (Wolpert, Lewis and Summerbell, 1975). The essence of their hypothesis is that the state of a cell determining its differentiation capacity changes progressively in time so long as it is in the progress zone, but once a cell leaves this zone its differentiation capacity rapidly becomes fixed or determined. Since there is a correlation between the time spent in the progress zone and the total number of division cycles which a cell undergoes, it is natural to identify the change of cell state with a quantal mitosis type of hypothesis, as is done by Wolpert *et al.* (1975). This must be modified to accommodate the observations made in Chapter 3 regarding Holtzer's ideas about mitosis and cell differentiation. Also, it is important to observe that cell divisions occur in the developing chick wing outside the progress zone, and differential growth is a significant aspect of limb morphogenesis. Thus some caution is required in making too simple a relationship between cell cycling and cell state in the chick limb.

Since the progress zone travels away from the embryonic flank, leaving behind it as it goes a trail of cells which have undergone progressively more cell divisions in the progress zone as distance from the flank increases, it is clear that the model generates a proximo-distal ordering of cell states which defines an axis of 'positional information' along the limb. No gradient of a diffusible substance is involved in such a process, cells simply 'remembering' their state when they leave the progress zone by some determination process. The difficulty which this model encounters is with evidence that the chick limb elements have some regulative capacity, since if cells become determined when they leave the progress zone then their fates are sealed. Wolpert *et al.* (1975) accommodate this by saying that there may be a small amount of cell–cell interaction outside the progress zone to allow for local adjustments, but they do not accept that much regulation is possible, despite evidence for this from the studies of Hampé (1959) and Kieny (1964).

The antero-posterior axis of the chick wing is determined by a region of cells at the posterior part of the distal limb bud known as the zone of polarizing activity (Saunders and Gosseling, 1968). If cells from this region are grafted to a more anterior position in the bud, then mirror symmetry of limb elements results. Thus one may regard them as the source of a substance which, by diffusion, establishes a gradient whose slope defines the antero-posterior axis and whose values define local positional information, as is done by Wolpert *et al.* (1975). One then has two quite different morphogenetic mechanisms operating in the chick wing: an epimorphic one involving growth together with a counting mechanism for the generation of the proximo-distal axis; and the familiar source-sink diffusion process for morphallaxis along the antero-posterior axis. Virtually nothing is known about the dorso-ventral axis.

The question naturally arises whether these apparently different types of morphogenetic process can be understood in terms of a single type of model. One is not, of course, compelled to adopt such a unified view even if it can be demonstrated, but considerations of simplicity and universality would favour it. Also, one may gain in conceptual clarity and versatility in adopting a model which applies equally well to both types of morphogenesis, for there is some ambiguity in the use of a concept such as positional information over a field whose boundaries are constantly changing. A somewhat more dynamic and flexible relationship between the field generating process and the domain being ordered seems desirable and this I believe is provided by the general principles of the wave model discussed in the last chapter. Let us now apply this to the establishment of the proximo-distal axis in the chick wing.

The Wave Model and Epimorphosis

Since the progress zone plays a special role in limb morphogenesis, it is natural to locate the origin of the propagating activity wave in this region of the developing limb. The wave will then propagate in a disto-proximal direction, the ionic and low molecular weight elements of the wave travelling from cell to cell via gap junctions and a gradient of bound morphogen will be set up with a maximum value in the progress zone. That this gives a reversed gradient profile compared with the model proposed by Wolpert *et al.* is of no consequence, since an axis is established by a gradient running in either direction. In order that the value of the morphogen in the progress zone should increase progressively with time, measured by cell divisions, we may assume that the wave is initiated once every cell cycle (or at some defined frequency per cell cycle) so that with each successive recurrence of the wave, the value of the membrane-bound morphogen increases. A saturation value will eventually be reached which is taken to be that determining the most distal elements of the wing.

To stop the rise in morphogen value in cells which have left the progress zone, it may be assumed that membrane excitability decays in cells outside this zone so that the wave fails to propagate further than some defined distance from the zone. The role of the apical ectodermal ridge is then to maintain the underlying mesenchyme cells in a state of continuous cell division and their membranes in a condition of excitability with respect to activity wave propagation, i.e. to sustain a state of high lability and activity characteristic of an organizer region. One then has the essential components of a model which has all the requisite properties for explaining the phenomena relating to proximo-distal axis formation in the chick wing. Whatever amount of regulation occurs can be explained in terms of cell excitability and the distance over which the wave can propagate. For example if, as a result of grafting buds to stumps at different ages, cells become excitable during the healing process so that waves can propagate beyond the normal limits outside the progress zone, then a smoothing of the gradient will occur and regulation will ensue.

The model makes a few suggestions regarding observations which could test its validity. Filming the development of the wing bud might reveal the occurrence of activity waves. Introducing at particular points in the limb bud a periodicity of a substance such as cAMP, which could be involved in gradient formation, might disturb normal morphogenesis. This would be the analogue of the pulser experiments of Robertson *et al.* (1972) on slime mould aggregation. Other agents such as potassium and calcium ionophores could interfere with the wave propagation process and disturb morphogenesis selectively. There is also the interesting possibility of carrying out electrophysiological studies in an attempt to observe spontaneous activity waves, despite the technical difficulties of such procedures.

Setting up a second axis by means of a propagating wave process is no more complex than establishing the first. One needs only a pacemaker region, which in the case of the antero-posterior axis of the chick wing is provided by the zone of polarizing activity, and a metabolic process on the membrane with the same general characteristics described in the last chapter. Some metabolites may be common to the activity waves generating different axes, in which case there will be definite patterns of interaction between the axes. These questions are discussed in somewhat more detail below in a consideration of the problems of neuro-embryology. It is sufficient to observe here that the apparent differences between epimorphic and morphallactic development do not necessitate different types of field-generating process, and that the wave model can readily be adapted to either case.

At this point let me recall to the reader's memory a remark made in Chapter 3 about the contrast between the mechanisms proposed in this book for setting up morphogenetic fields during embryogenesis and those assumed to operate in the maintenance of tissue homeostasis in the adult. Concerning the latter, a simple negative feed-back process involving diffusion of a control molecule from differentiating cells was employed. This was the chalone model. In seeking an understanding of morphogenetic field formation over a diverse range of organisms it was found in the last chapter that a rather more dynamic model was required, particularly one associated with membranes and so applicable to both unicellular and multicellular organisms. There is in fact no difficulty in reconciling the existence of these two processes in the metazoa, where control of cell division is an extremely important feature of organogenesis and differentiation. They work in a manner complementary to one another, the wave process generating global gradients which organize large-scale or field aspects of morphogenesis, while the local feed-back model stabilizes tissue architecture by regulating the processes of cell division and differentiation. One process can give way gradually to the other, the disappearance of the global gradient-generating process being perhaps the result of loss of cell excitability and hence loss of the capacity to propagate activity waves. Whether or not the tissue or organ is capable of regenerating then depends upon the details of its regenerative strategy. The difference between a rat and a newt with respect to limb regeneration may lie in a difference of excitability of cells in the stump and their ability to generate a morphogenetic field over the new tissue. The interesting studies of Becker (1972) on the capacity of small electrical currents to stimulate blastema formation and partial limb regeneration in rats are very suggestive in this connection. However, there appear to be situations in which limb regeneration can come about entirely as a result of local interaction rules between cells, old tissue specifying the boundary values of the field to be reconstructed by the growth of new tissue, in much the same manner as that proposed for skin regeneration. But a limb has three dimensional order and the skin only one, so it is necessary to consider how this higher ordering may come about.

Local and Global Order: the Imaginal Disc and the Cockroach Limb

The developmental system which I shall now consider is the insect limb. Embryologically this originates from the leg imaginal disc, which forms in the late embryo or early larva by thickening of the epidermis. During larval life the limb and other imaginal discs grow by cell division, attaining by invagination a characteristic sac-like organization in the mature larva. In the course of metamorphosis, the folded epithelium of the disc everts and the cells undergo changes of shape, differentiating into the components of the adult structure, a segmented appendage in the case of the limb.

If an imaginal disc dissected from a larva is cut in two and the pieces allowed to grow in an adult abdomen, one piece will regenerate in such a way that on being induced to differentiate in a metamorphosing host it produces all the structures of the normal appendage, while the other piece duplicates, producing a mirror-image of itself after differentiation. Detailed studies of this property in the leg disc by Schubiger (1971) and in the wing disc by Bryant (1975*a*) have established the behaviour to be expected from any piece of the disc. Bryant (1971, 1975*b*) interpreted this data in terms of a one-dimensional gradient of developmental capacity, but this fails to explain some of the observations, as pointed out by French, Bryant and Bryant (1976) in an analysis of pattern regulation in the *Drosophila* wing disc and in the cockroach and amphibian limbs. This is most readily demonstrated by reference to Figure 6.3. This shows the *Drosophila* wing disc and C is the

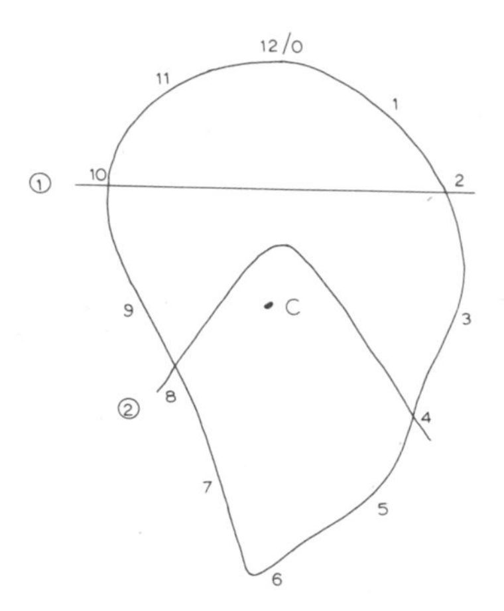

Figure 6.3. A schematic drawing of the *Drosophila* wing disc, with the circumferential values introduced by French *et al.* (1976) to explain the results of regeneration and duplication experiments.

point which Bryant (1975*a*) has deduced to be in some sense a peak of developmental capacity, since if the disc is cut into two pieces by a straight line such as ①, placed anywhere on the disc, the piece which contains C will regenerate and the piece without it will duplicate. However, a curved cut such as ② gives two pieces which fail to behave in the manner predicted by such a model, which requires that the piece containing C should regenerate. In fact, it duplicates, while the other piece regenerates. A formal model explaining this result is given in the paper by French, Bryant and Bryant (1976). This assumes that the disc is spatially organized according to a polar co-ordinate system with C as the origin and with circumferential values of an angular variable running around the periphery. They give these circumferential values the numbers 1 to 12, as in a clock-face, with the rule that 12 = 0 so that there is no discontinuity of value at the origin. The rules governing the behaviour of pieces of the disc are then easily stated. Wherever a cut is made, the exposed surfaces come together (in accordance with experimental evidence) and cell divisions occur along the line of union. New cells take on values which are the arithmetic means of their neighbours, calculated in accordance with the rule that these values are the ones which complete the missing numbers by the shortest possible route. Thus, considering the piece above cut ① in Figure 6.3, values 2 and 10 will come into coincidence after union along the cut and growth will occur between these values until the circumferential series 11, 12/0, 1 has been established. This obviously gives a duplication. The other piece will regenerate, since the shortest circumferential route between the values at the cut, 10 and 2, is 11, 12/0, 1 not 9, 8, 7, 6, 5, 4, 3. Thus the full set of circumferential values is restored and a normal structure will be formed. The values which cells take within the boundary follow, of course, the same averaging rule, using radial variables as well as the azimuthal values shown. The result is always a smooth two-dimensional map over the tissue, but in the duplicated cases the map is degenerate, two different points having the same value with mirror symmetry about a line.

In the case of the cut along the curve marked ②, one sees immediately that the rules stated above will give regeneration of the piece above the cut and duplication of the piece below, in accordance with observation, by virtue of the shortest distance procedure for re-establishing circumferential values. French and Bullière (1975*a,b*) had shown that these rules determine the regenerative behaviour of the cockroach limb in response to a great variety of cutting, grafting and rotation experiments, all of which are fully accounted for by the formalism. This is an exceedingly powerful and elegant result. It provides a simple, clear procedure whereby one can predict the outcome of any operation on the cockroach limb involving simple surgical manipulations. Since there is a great body of embryological literature which is devoted to just this type of operation, it will be of the greatest interest to see how far it can be comprehended within a single formal model of the type described above. The demonstration by French *et al.* (1976) that the rules applying to the cockroach limb also apply to *Drosophila* imaginal discs is a first step in this direction. However, the problem we now face is what type of morphogenetic field-generating process could give rise to a circumferential organization of the polar-co-ordinate type shown in Figure 6.3, together with the growth, averaging and shortest circumferential route rules which describe the experimental observations so elegantly. The following provides a possible solution to this interesting question.

Figure 6.5a shows schematically the circumference of a developing structure such as an imaginal disc, a cockroach limb, or an amphibian limb, represented in cross section. There are twelve co-ordinate positions labelled as on a clock-face, to facilitate comparison with the French *et al.* (1976) scheme, but now each co-ordinate position is assigned a pair of field values, $f_{(n)}$ and $g_{(n)}$, the subscript referring to the co-ordinate value. The field values $f_{(n)}$ form a gradient along the dorso-ventral axis, with symmetry anteriorly and posteriorly; while the other values form an antero-posterior gradient of a different substance or activity, with symmetry dorso-ventrally. This double gradient system conforms to a considerable body of classical work showing that the antero-posterior and the dorso-ventral axes are laid down separately at different times in developing limbs. This is true also of other organs such as the retina, to be discussed later. The difficulty with such a scheme is that, whereas it is easy to regenerate missing values between, say, 1 and 5 by

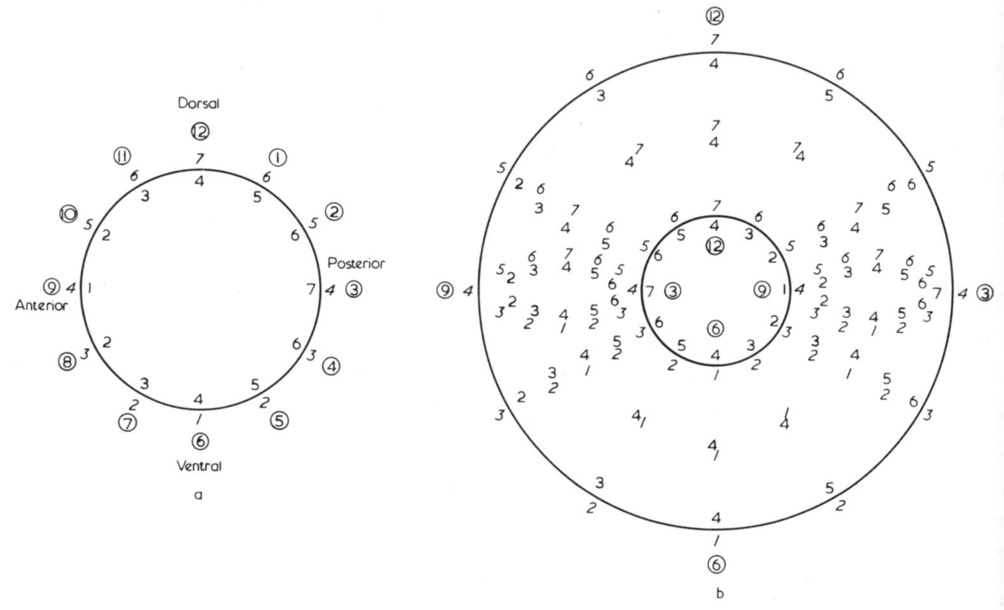

Figure 6.4. a. A schematic representation of the circumferential limb field, the dorso-ventral gradient of field values $f_{(n)}$ being shown in italic numbers, outside the circle, and the antero-posterior gradient $g_{(n)}$ in ordinary numbers inside.
b. The equilibrium field resulting from the application of the generative rules described in the text to the case of a graft of a right limb (inner circle) onto a left stump (outer circle), showing the positions and orientations of the supernumerary limbs.

an averaging process which continues until all the missing values are recovered, it is not obvious how to regenerate an extremum (a 1 or a 7) if it is removed by a cut. Such extrema act like boundary values, and so far no model has been advanced which gives a prescription for both the regeneration of boundary values and the smoothing (averaging) of intermediate field values. It is, however, fairly obvious that unless these two processes can be accounted for, one does not have a model of developmental field dynamics. Simple rules which give this behaviour will now be stated. They derive from studies carried out in collaboration with Drs F. W. Cummings and J. W. Prothero. A complete description of this extended field approach, giving rules for boundary value and internal field value specification for a variety of pattern-forming processes including hydroid regeneration, somitogenesis, feather pattern formation, phyllotaxis and limb regeneration, is in preparation for publication.

Consider first the process of intercalary growth, i.e. the appearance of new tissue when two cut surfaces with different field values are brought together. An explicit statement of this process, in terms of the field numbers of Figure 6.4a, is that growth occurs at a discontinuity of the field whenever the slope of either set of field values is greater than 1. (The slope of the field f at co-ordinate position n is defined as

$$Df_{(n)} = \frac{f_{n+1} - f_{n-1}}{2},$$

and similarly for g. Co-ordinates are numbered modulo 12.) Field values are assigned to the new tissue in accordance with the following rules, whose significance will become evident shortly.

If $Df_{(n)} \geqslant 1$, then the new field value at co-ordinate position n is

$$f'_{(n)} = f_{(n)} - C(f_{(n)} - \bar{f}_{(n)}) \tag{6.2}$$

where $f'_{(n)}$ means the new value to be assigned to tissue in position n, the previous value being $f_{(n)}$; C is a constant; and $\bar{f}_{(n)}$ is the spatial average,

$$\frac{f_{n-1} + f_{n+1}}{2}.$$

This is simply an averaging rule. If $Df_{(n)} < 1$, then

$$f'_{(n)} = f_{(n)} + K(f_{(n)} - 4) \tag{6.3}$$

where K is a constant. This has the consequence that if $f_{(n)} > 4$, then $f'_{(n)} > f_{(n)}$; while if $f_{(n)} < 4$, then $f'_{(n)} < f_{(n)}$. There is thus a progressive divergence of field values until either the maximum (7) or the minimum (1) is reached. Identical rules apply to the other values, g. Therefore the slope determines whether new tissue takes field values according to a divergence rule, generating an extremum (maximum or minimum), or according to an averaging rule. Speaking metaphorically, these rules imply that cells 'like' to be in a developmental field of slope 1, smoothing out values if the slope is greater than this and generating a greater slope if it is less than 1. Of course the numbers in this representation are purely arbitrary. The value 4 in expression (6.3) is a constant parameter, genetically determined.

Let us see how these rules generate the required results for various operations. It is assumed that the field values of old tissue remain unchanged. Suppose a cut is made between co-ordinate positions ⑥ and ⑦, and between ⑨ and ⑩, dividing the structure into two parts. Considering first the smaller part, field values (*2*, 3) and (*4*, 1) will be brought together at the cut surfaces. The slopes are greater than 1, so the averaging mode (6.2) is used to assign field values to the new tissue. Taking $C = 1$ for convenience, we find that the new field values are

$$\tfrac{1}{2}(f_{⑦} + f_{⑨}) = \tfrac{1}{2}(2 + 4) = 3$$

and

$$\tfrac{1}{2}(g_{⑦} + g_{⑨}) = \tfrac{1}{2}(3 + 1) = 2$$

These values are stable, so the result is a duplication of this smaller part, as observed experimentally.

In the larger part, field values (*1*, 4) and (*5*, 2) will come into juxtaposition. Growth then occurs and the new tissue at co-ordinate position 8 is assigned field values by averaging, giving (*3*, 3). The slope in f is still greater than 1 between the new and the old tissue, so further growth occurs and averaging continues. The new values at positions ⑦ and ⑨ are, therefore

$$f_{⑦} = \tfrac{1}{2}(f_{⑥} + f_{⑧}) = \tfrac{1}{2}(1 + 3) = 2$$

$$f_{⑨} = \tfrac{1}{2}(f_{⑧} + f_{⑩}) = \tfrac{1}{2}(3 + 5) = 4.$$

This gives the series *2, 3, 4* in the regenerate, which is stable. The other field values are

$$g_{⑦} = \tfrac{1}{2}(g_{⑥} + g_{⑧}) = \tfrac{1}{2}(4 + 3) = 3{\cdot}5$$

$$g_{⑨} = \tfrac{1}{2}(g_{⑧} + g_{⑩}) = \tfrac{1}{2}(3 + 2) = 2{\cdot}5$$

The series in this field is thus 2·5, 3, 3·5 with slopes less than 1 at each position. Therefore the divergent mode (6.3) is adopted. We take $K = 1$ for convenience, and get the results

$$g'_{⑦} = 3{\cdot}5 + (3{\cdot}5 - 4) = 3$$

$$g'_{⑧} = 3{\cdot}0 + (3{\cdot}0 - 4) = 2$$

$$g'_{⑨} = 2{\cdot}5 + (2{\cdot}5 - 4) = 1$$

This is now a stable series as well and the overall result is the regeneration of the missing values of the larger part, as required. One sees in this example how the divergent mode generates an extremum (in this case, a minimum). If the cut had removed a maximum, it would have been regenerated by the same rule.

Consider next the type of grafting experiment which gives rise to supernumerary limbs, the prediction of which gives considerable support to the formal scheme advanced by French *et al.* If a regenerated blastema in an amphibian, or a cockroach limb, is grafted to a stump of the opposite limb such that the dorso-ventral axes are aligned but the antero-posterior axes are in opposition, the result is a graft from which arise two supernumerary limbs with the same antero-posterior polarity as the stump. This situation is shown schematically in Figure 6.4b, where the outer circle represents the stump and the inner the graft. The generative rules for field value formation have been applied to give the number sequences as shown, producing the required results, since wherever a complete circumferential series of field values arises, it is assumed that there a new limb will form. For example, between stump co-ordinate position ② and graft position ⑩, field values (*5*, 6) and (*5*, 2) confront one another. The discontinuity between 6 and 2 results in intercalary growth and the assignment of field values to the new tissue in accordance with the averaging rule, giving the stable series 6, 5, 4, 3, 2. The other field values are *5*, *5*, with slope less than 1, so the divergent mode is used and the series generated is *5, 6, 7, 6, 5,* regenerating the maximum. This is true of all the field series $f_{ⓝ}$ in the upper half of the graft, while the other extremal value, 1, is generated throughout the lower half. At the points where field values 4, 4 are juxtaposed there is a condition of unstable equilibrium, new field values being the same as the old: $f'_{③} = 4 + (4 - 4) = 4$. However, the slightest perturbation of these values above *4* results in divergence to *7*, and below *4*, to *1*. Since tissue above the horizontal diverges to *7*, it is natural to assume that this perturbs adjacent tissue and the series *4, 5, 6, 7, 6, 5, 4* is produced; but below the horizontal, divergence proceeds in the opposite direction, giving the series *4, 3, 2, 1, 2, 3, 4.* Thus the circumferential sequence is generated, giving the required supernumeraries.

In applying these rules to other graft combinations, it is important to beware of artefacts arising from the coarseness of the numbering series and the consequences of choosing parameters such as $C = K = 1$ in expressions (6.2) and (6.3). These were chosen for simplicity of demonstration. In reality, field values will naturally be more continuous, a representation using numbers 1 to 70 being more plausible than 1 to 7; while the parameters must be chosen to take account of experimental behaviour.

There are certain properties of the scheme outlined above which are worth commenting on. A double-gradient system is the obvious way of ordering a circumferential field and is the solution which has been most widely adopted. However, to reproduce the experimental observations, rules whereby extremal field values are regenerated need to be introduced. These require physiological interpretation. The switch between divergent and convergent (averaging) behaviour in response to field slope may be seen as a membrane-dependent process involving a phase transition between conditions of excitability and non-excitability of the type analysed in the next chapter in relation to the membrane wave model, resulting in gradient-generating and averaging behaviour, respectively. The nature of the field slope measuring process is also obscure, but the exploration of neighbouring cell surfaces by filopodia could result in the generation of forces across cells which operate such switches. Although the generative rules suggested are by no means unique, it appears that some kind of cell surface effect connected to a divergence/convergence selector is required for circumferential field organization. Knowing what qualitative behaviour is required should help to focus attention on the nature of the physiological processes involved at the cellular level.

It is evident from these examples, that it is necessary to make a clear distinction between the notion of co-ordinate value, designating position in a tissue according to some convenient scheme, and field value, referring to cell or cortical state. The former is fixed by some convention, but the latter changes according to rules or dynamical constraints. The use of the term 'positional value' confuses these and results in ambiguity.

It is of interest to observe that the averaging expression (6.2) is a discrete version of the equation

$$\frac{\partial V}{\partial t} = \nabla^2 V(x, t)$$

whose time-independent solutions are those of Laplace's equation,

$$\nabla^2 V(x) = 0, \quad V(x)$$

being the field value at position x. Thus we are dealing with a field with the same type of behaviour as a simple diffusion process. However, there is no need to interpret the underlying mechanisms as those of chemical diffusion. One is thus freed from the constraint of representing developmental fields in terms of diffusion-reaction schemes, resulting in a great simplification of the analytical description of field-generating processes. Diffusion-reaction systems continue to be of use in the detailed study of particular processes, but their complexity and dependence upon arbitrary parameters make them unsuitable for a general theory of development. The divergent mode, (6.3), extends the classical field description to include the generation of boundary or extremal values. Thus the model suggested involves a generalized field theory of extreme simplicity and considerable descriptive power.

THE RETINO–TECTAL PROJECTION OF THE AMPHIBIAN VISUAL SYSTEM

Let us now turn our attention to a developmental process which involves not simply the ordering of a two-dimensional array of cells, but the mapping of one two-dimensional surface onto another, a process which has attracted the attention of a number of highly competent and imaginative investigators and hence for which there exists a very interesting body of data. This is the outgrowth of optic fibres from the ganglion cell layer of the retina, their passage together out of the optic stalk as the optic nerve, growth to the contralateral optic tectum and orderly establishment of connections with neurons in the tectum. The essential question posed by many investigators of this phenomenon is how it is that the retinal fibres making up the optic nerve find topographically correct positions on the optic tectum and there establish functional connections so that photic stimulation of a particular region of the retina gives rise to excitation of neurons in defined regions of the optic tectum. This mapping process has been studied by a variety of techniques which include surgery on the retina or tectum or both, followed by behavioural, histological or electrophysiological analysis. Pieces of retina or tectum can be removed, rotated, fused with parts of other retinae, the optic nerve can be induced to arrive at the tectum by an alternative route or sent to the ipsilateral tectum, or combinations of these procedures can be followed. The resulting projection of the retinal neurons on the tectum is then observed by some means, the most satisfactory being electrophysiological recording of the responses of the tectal surface to stimulation of the retina by a point of light, and conclusions are then drawn from any significant perturbations from the normal map. Early studies by Stone (1948) and Szekely (1957) by surgery and behavioural analysis established that a rotation of the eye at an embryonic age later than Stage 36 in *Ambystoma* or Stage 22 in *Triturus* resulted in an equivalent rotation of the visual field as observed by behavioural analysis. Thus if the left eye of a Stage 36 salamander embryo is rotated clockwise through 90°, then after metamorphosis the adult will snap in a forward direction at a fly or a lure presented in his left visual field and will never learn to correct this behavioural defect. However, if the rotation is made at a developmental age earlier than Stage 34, the behaviour of the adult is perfectly normal. Szekely's studies showed further that the naso-temporal (antero-posterior) axis of the eye in *Triturus* is fixed before the dorso-ventral axis. Jacobson (1968*a*) working with *Xenopus*, confirmed that the projection as studied electrophysiologically validates this picture: if a 180° rotation is made in a Stage 30 *Xenopus* embryo, then one axis (naso-temporal) is reversed but the other axis is normal. This tells us that the axes within the retina become fixed or determined at defined developmental ages. Jacobson (1968*b*) also established a correlation between this fixation and the cessation of DNA replication in the central region of the retina, suggesting that there may be a causal relationship between field lability and the cell cycle of the same kind as that observed in the insect epidermis. Since at the stage of fixation of axes the retina is very much smaller than its final size, growth occurring by addition of cells to the periphery (ciliary margin), the rotation experiments show that after Stage 32 new retinal cells take their morphogenetic information from the determined field within the retina, not from the body axes. It is possible that the retina loses its connections with surrounding cells at this stage and so becomes autonomous. In a dynamic process such as the propagating wave model, the extension of the maximum of a gradient into cells added to a field by growth occurs naturally, since the high point of the gradient will extend to newly added cells in view of the fact that membrane which is shared by daughter cells contains the morphogen and continuous wave activity will keep pushing the gradient peak into the new cells. Such behaviour is also characteristic of a polar-pumping model of the type proposed by Wilby and Webster (1970), but this mechanism encounters difficulties in more than one dimension due to the problem of directed pumping of different substances at right angles to one another through the same cell.

The absence of regulative capacity in the visual system of adult amphibia, first demonstrated by Sperry (1944), suggested some rigid mechanism whereby optic fibres make connection with tectal neurons once the axes in the two tissues have been fixed. Sperry (1951, 1955, 1963) proposed that this was achieved by the specific complementary labelling of the cell surfaces of retinal and tectal neurons according to their positions in the tissue so that unique functional connections were established by retinal fibres with appropriate tectal cells. The hypothesis has been called 'neuronal specificity'. A great deal of work using electrophysiological procedures to map the retino-tectal projection has been devoted to the investigation of this hypothesis. The resolution of these techniques, though high by embryological standards, does not allow one to make any conclusions about the specificity of cell-cell connections to a discrimination level of less than about 1000 cells. Excision and reimplantation studies of Levine and Jacobson (1974) on the optic tectum have, however, suggested a very much finer target accuracy of retinal fibres, down to 5 cell diameters or so (Jacobson, 1975). It must be remembered that upon reaching its destination on the tectum, an optic fibre branches extensively and so makes functional contact with many tectal neurons. Furthermore, many ganglion cells in the retina will be in functional contact through their fibres with many tectal neurons. The evidence which has accumulated largely through the work of R. M. Gaze and his collaborators [Gaze, Jacobson and Szekely (1963, 1965); Gaze, Keating and Straznicky (1970); Gaze and Keating (1972)], strongly suggests that the interactions between retinal fibres and tectal neurons are not specific in the sense that Sperry originally suggested, but are such that

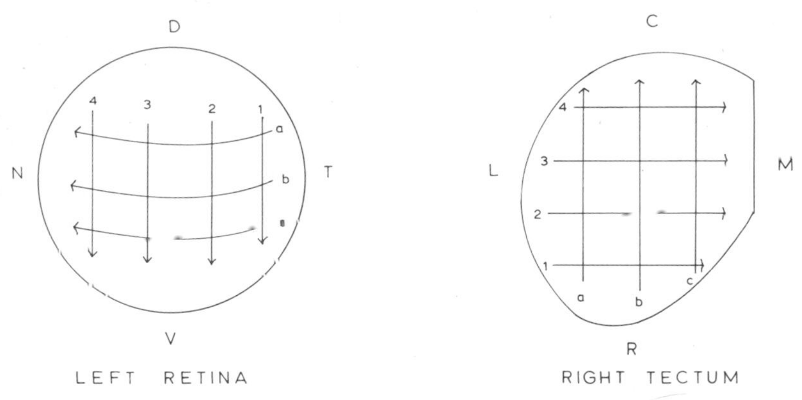

Figure 6.5. A diagram of the retino-tectal map in *Xenopus laevis.*

relative ordering only of the fibres over the two axes of the tectum is achieved. The essential features of the retino-tectal projection are shown in Figure 6.5.

In constructing the map electrophysiologically, an external electrode is placed at a marked grid position on the tectum. A spot of light is then moved over the retina, the light being located within a frame representing the visual field and the point of the retina giving maximal response on the tectum is recorded as the corresponding map position. Since the tectal surface curves ventrally in the lateral region, lateral points are not easily reached by an electrode which is positioned from the dorsal direction, the tectum being exposed by removal of the dorsal part of the skull. Therefore these points are not usually mapped and so there is a region of the visual field which remains empty of projection points. This is the inferior visual field, which maps to the lateral tectum. Because of the camera inversion effect of the lens, inferior visual field points fall upon dorsal parts of the retina. Thus it is the dorsal retina which is not usually mapped in electrophysiological studies, as seen in Figure 6.5. However, it is clear from studies in which the lateral tectum has been mapped by special procedures that the projection is a continuous one. In the literature one usually finds maps of the visual fields to the optic tectum, since this is how the experiments are carried out. However, Figure 6.5 shows the map in terms of the retina and the tectum, since our primary interest is with the epigenetic relationships between these two tissues.

It is interesting to note that the arrows labelled a, b and c on retina and tectum, denoting the axes which run parallel to the antero-posterior axis of the embryo (the eyes are located at the side of the head in *Xenopus* embryos and there is practically no binocular overlap), have their arrowheads pointing in opposite directions relative to the body axis. Some insight into how this might come about is suggested by the studies of C. O. Jacobson (1959) on the presumptive cerebral regions in the neural plate of the axolotl. Here it is shown that the eye primordia, connected by the presumptive optic chiasma, lie anterior to the presumptive optic tecta. A pacemaker centre located in the anterior-central region of the neural plate at Stage 14–15, between the presumptive diencephalon and mesencephalon, would in fact establish gradients over these regions which run in opposite directions when considering eye and tectal primordia in contralateral pairs. Such axes would not run parallel to the main body axis, but the morphogenetic movements which accompany the development of the retina and the tectum from the primordia could bring about such an alignment. As regards the other direction, it is not clear whether both tissues are specified by a single body axis, since the second axis in the retina is primarily dorso-ventral while that in the tectum is primarily medio-lateral. However, on a curved surface such as the embryo, medio-lateral and dorso-ventral may be part of the same axis. Such problems can only be resolved by transplantation experiments of retinae into tissues with distinct medio-lateral orientations and seeing what axial specifications result.

In *Xenopus laevis*, the adult optic nerve consists of about 50 000 fibres arising from ganglion cells in the retina which map on to an approximately equal number of neurons in the contralateral optic tectum, complete crossing-over or decussation occurring at the optic chiasma. The nerves must grow from the retinae to the ventral surface of the thalamus, cross over the mid-line and so establish a chiasma and then grow dorsally towards the respective tecta. The nerves split into medial and lateral tracts just prior to reaching the rostral part of the tectum and these branches carry fibres appropriate to the tectal regions to which they grow. Thus the fibres are already sorted out in the optic nerve and the branching is in accordance with their destinations. Evidently these fibres know very well what they are about. The crossing of the mid-line and the formation of the optic chiasma is itself a very interesting problem of directed growth and the splitting into two branches even more intriguing, but these are questions which I do not intend to go into here. Let me just observe that the optic nerve can be deflected from its normal course and induced to grow to the caudal tectal surface by the oculomotor route, after which the fibres map in correct topographic order across the tectum. So we cannot use the normal rostral arrival of retinal fibres at the tectum as an important factor in the mapping process; nor can we use developmental age as an element in the mapping, since it occurs perfectly normally when optic nerves are cut and allowed to regenerate in the adult. Finally, let us observe that an optic nerve can be deflected to the ipsilateral tectum, where the fibres will project normally, so that there is no intrinsic information which distinguishes right from left. This makes the normal chiasma crossing process somewhat more puzzling, but let us concentrate now on the process of retinal fibre mapping over the tectal surface.

A MECHANISM FOR TWO–DIMENSIONAL PROJECTIVE ORDERING

Let us suppose that along the primary axes of retina and tectum, viz., naso-temporal and rostro-caudal, respectively, there are complementary gradients of adhesive markers and adhesiveness. Thus the membrane-bound morphogen gradients along these axes (call them M_1^R and M_1^T for retina and tectum respectively), could induce the incorporation of glycoproteins or other markers into the cell surfaces so that there are graded distributions of these molecules over the retina and the tectum. We assume that on the tectum maximum adhesiveness occurs rostrally, while in the retina, the maximum of adhesive markers (which may be different from those of the tectum) occurs at the temporal extremity. We assume that the fibres growing out from the retinal neurons carry these markers in the same relative concentrations as their cell bodies. Such graded distributions will give a one-dimensional axis of order between the two surfaces. What about the second?

Let us suppose that along the second axes of the retina and tectum, the gradients of the second morphogens, M_2^R and M_2^T, result in the occurrence of a spontaneous frequency of membrane depolarization which has the effect of inducing conformational changes such that marker proteins at the cell surface are much less exposed than when there is no membrane depolarization. Thus there will be an oscillation in the adhesiveness of a retinal fibre at a frequency characteristic of its position along the dorso-ventral axis, the frequency being maximal ventrally, say. We assume that there is a corresponding frequency gradient along the medio-lateral axis of the tectum, with a maximum medially. These properties will allow for the mapping of retinal fibres on to tectal neurons and the establishment of maximally-stable contacts when the two gradients are correctly ordered relative to one another. Furthermore, the ordering is relative, not absolute, as we shall now see.

Consider what happens when an optic nerve reaches a tectum by either the normal or by some other route, and in either an embryo or in an adult undergoing nerve regeneration. The filopodia extending out from the growth cone of a retinal fibre will explore the tectal surface over a distance of 100 μm or more and form contacts with any available adhesive sites. There is a spontaneous mobility of extension and retraction of filopodia, so that the fibres will tend always to move in a direction which allows for the formation of the greatest number of contacts between filopodia and tectal neurons. This will occur where fibres and tectal neurons have the same frequency and phase, and where potential adhesive contacts are greatest. Phase coincidence arises as a consequence of spontaneous phase variations between cells which result in a mutual drift until phases as well as frequencies coincide and allow maximum adhesive contact to occur along the medio-lateral axis of the tectum. Along the rostro-caudal axis fibres will jostle one another, filopodia making and breaking contacts with tectal neurons and competing for available adhesive sites. Those fibres with highest density of sites will tend to establish strongest contacts with rostral tectal neurons, which have the highest density of complementary sites, displacing filopodia which make weaker contact. There will be a pushing and shoving of fibres along both axes with a resultant tendency for them to spread over the whole available surface. The projection thus obtained is not an absolute one, since any retinal fibre can make adhesive connections with any region of the tectum rostro-caudally, while along the medio-lateral axis the map is relative also in view of the fact that some adhesion will in general be possible even without frequency matching. So it is the competitive interactions between retinal fibres as well as the co-operative interactions with tectal neurons that results in the mapping over the whole of the available tectal surface. There is a considerable amount of experimental evidence, persuasively presented by Gaze and Keating (1972), which argues strongly in favour of this kind of relative mapping or 'systems matching' process in the visual system. The particular model presented above is but one realization of a whole class of projection mechanisms with this characteristic. The description has been entirely verbal, and requires mathematical and computational verification.

Up to now we have ignored the fact that the retino-tectal projection has a third dimension of order, for on reaching their correct destinations retinal fibres plunge into the tectum and ordered branching occurs within the depth dimension. To account for such behaviour we may assume that upon achieving frequency and phase coincidence and having established stable contacts with adhesive markers on tectal neurons, the growth cone, now relatively firmly anchored on the tectal surface, begins spontaneously to explore the third dimension of the tectum. Gap junctions may also be formed as a result of stable surface contacts so that retinal fibres and tectal neurons begin to interact dynamically. Since ultimately it will be the job of retinal fibres to relay trains of action potentials to the tectal neurons in response to sensory stimulation of the retina, functional contact must be established. But since so little experimental evidence is available about the nature of the depth ordering in the tectum, it is difficult to know exactly what epigenetic problems need to be resolved. Suffice it to say at the moment that the model described has characteristics which satisfy both the general requirements of the retino-tectal projection process and also anticipate specifically neural features of the system, such as the occurrence of spontaneous frequencies which are used in the mapping process itself. These frequencies I imagine to be low relative to the frequencies of discharge of mature neurons, which are of the order of several per second, since we are still dealing with developmental phenomena involving metabolic processes. They would tend rather to fall into the time scale of events associated with intermediate memory; and in view of their relationship to a mapping process which involves the establishment of neural contacts and functional synapses, they could well be fairly directly connected with such processes in the mature nervous system as a continuation and specialization of an embryological mechanism.

The model presented here for two-dimensional order in a sheet of cells is not by any means intended as a solution of the retino-tectal mapping problem, still less as a general solution of the problems of cell sorting in two dimensions. It is intended to illustrate the nature of these problems, and to suggest how dynamic (oscillatory) behaviour of membranes can be used as an alternative to a static adhesion gradient. The question regarding the specificity of adhesive markers has not been considered, although evidence by Barbera, Marchase and Roth (1973) suggests that there are qualitatively different adhesive sites on the medial and lateral tectum of the chick, with complementary sites on ventral and dorsal neural retina cells. There is as yet no similar evidence for the amphibian eye and brain. Such specificity may arise from a double gradient of markers along these axes, a situation which could result from a single initial morphogen gradient which activates one type of enzyme involved in surface marker synthesis and inhibits another. A double gradient of specific adhesive sites of this kind would again produce a relative, not an absolute, map, as shown by Prestige and Willshaw (1975). Referring back briefly to the observations of Bohn (1974), Lawrence (1974) and Nübler-Jung (1974) on cell movement in the insect epidermis, one can see how this could come about by the same principles of filopod formation and exploration of maximally stable adhesive contacts as that occurring in the retino-tectal mapping process, but there the problem is restricted to a single population of cells.

Some evidence for a highly plastic, dynamic character of the tectal map is provided by recent experiments of Chung and Cooke (1975). They studied the retino-tectal projections resulting from 180° rotations of presumptive mid-brain regions in *Xenopus* embyros, Stages 21–24 and 37, and found evidence that a region in the diencephalon acts as an organizer for the tectal map, determining its polarity. The implication is that up to Stage 37, tectal cells do not acquire fixed positional markers. This is consistent with either a phase-shift type of model, or with the wave model discussed above with the modification that the pacemaker lies in diencephalic tissue and continues to be active until Stage 37. Rotation and transplantation studies on adult brain tissue by Yoon (1973), by Levine and Jacobson (1974) and by Jacobson and Levine (1975) give an equivocal picture of the capacity of differentiated brain cells to remember their previous positions, so the status of positional labels in the adult brain is not clear. Evidently one needs to retain a flexible view on this matter. The dynamic models discussed above allow for a transition from plastic to fixed at any stage of development.

As a final remark, I should like to hazard the guess that over the next few years developmental biology will witness the emergence of new dynamical models and experimental observations in which periodicities and waves will feature very strongly. The notion of static diffusion gradients ordering spatial domains has served embryologists very well for a great many years, but such models simply do not provide the plasticity and the dynamic versatility required for morphogenesis and pattern formation. Activity waves in developing organisms can have a considerable range of velocities depending upon whether the processes involved are primarily electrochemical, metabolic or epigenetic. Here we encounter once again the hierarchy of relaxation times which we studied in the first two chapters of the book, but now we have added the spatial dimension to the temporal. The organism now begins to emerge as a system whose spatial and temporal order arises as a result of wave propagation from defined organizing centres, together with clocks (local or global) which generate spatial and temporal periodicities. This dynamic picture will involve the replacement of the relatively artificial positional information/interpretation dichotomy by a more continuous model of spatio-temporal unfolding, with critical discontinuities occurring as part of the ordered emergence of structure and form. How this is to be handled analytically is in part the concern of the next chapter.

Chapter 7

THE ORGANISM AS A COGNITIVE AND CO-OPERATIVE SYSTEM

IN this last chapter I shall explore an approach to the analysis of cells, embryos, and organisms in general, which may unify and extend some of the ideas which have been developed so far. The context for this exploration is the view that organisms are essentially cognitive and co-operative systems: cognitive in so far as they function and evolve on the basis of knowledge of themselves and their environment, co-operative in so far as the dynamical modes of operation of biological systems (growth, differentiation, rhythmic activity, etc.) are to be understood in terms of the co-operative (correlated) behaviour of particular processes. The second half of this view is relatively uncontentious and is simply a recognition that processes analogous to phase transitions underlie the rather sharp changes of state which biological systems undergo not just during embryological development, where they are much in evidence, but as an important aspect of any biological process. Thus the transition from the non-growing to the growing condition in a cell, the appearance of excitability in a membrane, the emergence of a pacemaker centre in a developing organism, may all be described as the development of a particular type of collective order in a system and examples of such processes will be treated analytically. They arise as fairly natural extensions of the ideas developed in previous chapters.

However, the view of organisms as cognitive systems does not automatically arise from the analysis developed thus far. I shall show that it is completely consistent with this analysis and is in certain respects a consequence of particular assumptions about biological organization which have been tacit in the previous treatment. Nothing has so far been said about the nature of biological processes which induces us to recognize their difference from the processes of inanimate nature, although a distinction has been implicitly assumed. All the models introduced are recognizably chemical and physical, but they are assumed to function in a biological context. It is necessary now to be somewhat more explicit about that context. At the higher levels of biological order, there is no difficulty in identifying characteristics which distinguish living from non-living systems: complex, purposeful behaviour patterns, the capacity to learn, the use of symbol-systems such as language, etc. In choosing to characterize organisms as cognitive systems,

I am deliberately selecting a high-level property for my definition of the domain of biological process. In doing so I believe that I am simply extending in a fairly obvious manner a viewpoint which has been developing with increasing momentum during the past decade or so.

Consider, for example, the position adopted by Chomsky (1968, 1972) in his analysis of language, particularly as revealed by the concept of linguistic competence. This implies an instinctive, unlearned capacity for generating correct sentence structure or syntax, a capacity which emerges in the course of the human developmental process. The rules and constraints which constitute linguistic competence define the processes which generate the surface structure of sentences from their deep structure, such as structure-dependent operations in sentence transformation. Possession of these rules, i.e., possession of the structural (anatomical) and functional (physiological) constraints which are the embodiments of the rules, is equivalent to having the knowledge required for speaking correct sentences. This knowledge is not learned, but is innate, inherited as part of the human phenotype. Thus the proposition that there are innate structures in man which endow him with linguistic competence is a statement that a certain type of knowledge is inherited. The more contentious part of this assertion is not that linguistic competence is inherited; it is that the structures and functions embodying the rules for generating correct speech constitute knowledge. Chomsky (1972) goes so far as to say that 'Knowledge of language results from the interplay of initially given structures of mind, maturational processes and interaction with the environment'. Innate structures are thus seen to constitute elements of knowledge. I shall simply use this proposition in a more extended form and suggest that the basic attribute of living organisms is their possession of knowledge about aspects of the world, knowledge which renders them competent to survive and reproduce in the environment to which they are adapted or which they know.

There are several reasons for adopting this viewpoint. The first is that there has developed in recent years a tendency for ideas from linguistics, visual scene analysis and the area of artificial intelligence in general, to be applied to biological processes in a relatively loose, analogical manner. I believe that comparative explorations of this kind are important and will be fruitful, but that they need to be conducted with some rigour if the comparisons are to be illuminating rather than confusing. A primary requirement for clarity is to have a working definition of knowledge which applies to higher cognitive activities as well as to biological process. I shall attempt to provide this and to explore some of its implications.

A second reason for this procedure follows from the first, which is that the domains of the biological and the human sciences may be comprehended within a single framework of ideas. In uniting these domains within a cognitive framework, no reduction of one level to another results. Important distinctions between, for example, the psychological and the physiological domains persist, but instead of being defined generally in terms of knowledge they are described in terms of such concepts as the capacity to learn, the type of learning involved and the time-scale over which this takes place. Thus I shall take the view, by no means original, that the evolutionary process involves a type of learning, but that it employs different mechanisms from those involved in the learning process in organisms, being both more wasteful and much slower.

A third justification for a cognitive view of biological process is the provision of a clear distinction between animate and inanimate modes of organization, the living and the non-living domains. This distinction persists between man and his artefacts, including computers and programs, which turn out not to be cognitive systems in terms of the definition advanced below. This is because the knowledge they undoubtedly contain is not self-referential: it does not contribute to their own stability in the world (their capacity to survive, reproduce and evolve).

Yet another consequence of this approach to organismic order is the possibility of a different attitude to the evolutionary process, and to Nature as a whole. This will be seen as a process manifesting intelligence, organisms being the results of the working of an intelligent system. This proposition may sound naively anthropomorphic to those wishing to pursue a reductionist approach in biology. It stems directly from the adoption of a high-level concept for the definition of biological process, and provides what I consider to be a self-consistent and rigorous alternative to a reductionist position, which encounters its greatest difficulties in the higher cognitive domains. Just the reverse is true of the present view, but I shall attempt to demonstrate its validity.

The connection between the cognitive and the co-operative aspects of the analysis arises from the fact that the dynamic modes which express evolved strategies of organismic behaviour are discontinuous one from the other, so that one needs a theory of bifurcation to describe them. The appearance of such discrete modes is most readily understood in terms of co-operative or collective behaviour, with the emergence of qualitatively different properties in the system beyond the bifurcation point. The different dynamic modes manifest different strategies of behaviour, different aspects of the organism's knowledge about the world.

The general plan of the chapter is to pursue these arguments in relation to specific mathematical models which provide rigorous frameworks for the propositions. However, the models are suggestive rather than definitive. Thus the chapter takes the form of a sketch, a plan for further exploration.

THE FUNCTIONAL HIERARCHY

A major theme in the analysis of biological systems presented in this book has been the identification of levels of organization in terms of the relaxation time of processes involved in the manifestation of a particular type of behaviour. Thus in Chapter 1 the metabolic system of a cell was identified by the relaxation time of enzyme-catalysed metabolic transformations in which enzyme concentrations were taken to be constant. The epigenetic system was then identified in Chapter 2 in relation to the characteristic time for macromolecular syntheses and interactions; while in Chapter 3 the genetic system was defined, although no analysis of its behaviour was presented. Such an analysis could be extended in either direction. Beyond the metabolic system, for example, there is what we could refer to as the electrochemical or the ionic system. This is primarily associated with membranes and involves ion fluxes such as those which accompany the action potential of the nerve axon. This has a relaxation time of the order of a few milliseconds. The close association of such events with energy-dependent metabolic processes such as membrane-bound ATPase activity makes an exact temporal separation of electrochemical and metabolic processes somewhat artificial in certain circumstances, but in general such a distinction is a useful one and serves to identify the difference of time scale which separates processes studied by neurophysiologists from those studied by, for example, biochemists. Furthermore, although our attention has been primarily restricted to simple organisms, the physiology of a complex metazoan requires a division of the epigenetic system into sub-systems in order properly to reflect the differences of time-scale on which such organisms operate. We have only to consider such obvious examples as tidal, diurnal, weekly, lunar and seasonal 'clocks' to recognize this spectral range (Bünning, 1967; Iberall, 1969; Richter, 1965), not to mention other processes such as regeneration, wound-healing, muscular adaptation, etc.

This analysis of organismic order into a functional hierarchy which involves the parametric control of one level of behaviour by processes at the next level, as described specifically in Chapter 4 in reference to the chemostat as an adaptive system, can be interpreted in relation to the concept of a hierarchy of operators, each carrying out observations by time-averaging over periods considerably longer than the relaxation time of the system 'below' it and initiating responses appropriate to the observations. Such a viewpoint is implicit in the analysis presented in previous chapters. For example, slow changes in the epigenetic system of a bacterium which result from prolonged changes of state in the metabolic system (e.g. derepression of the arginine biosynthetic enzymes as a result of a reduced level or exogenous arginine) occur as a result of the action of highly specific macromolecular 'observers' (e.g. permeases, repressors, polymerases, etc.). The population of observers involved in the control of such a physiological circuit has the effect of regulating the rate of particular processes, gene transcription being a primary one in this case, in accordance with the values of particular metabolic concentrations (inducers, co-repressors, etc.) averaged over a period of many minutes. Fluctuations of metabolite level with periods of 10^2 s or less are not reflected in the dynamics of the epigenetic system, whose operating time is many minutes up to an hour or so in bacteria. Let us see how we can describe the operation of such biological control processes in cognitive terms.

A cognitive system is one which operates on the basis of knowledge of itself and its environment. Knowledge I will take to mean a useful description of some aspect of the world (system or environment). The fact that we are dealing with a description means that there is some code or set of codes which relates it to that which is described; while usefulness in this context is connected specifically with the capacity to survive, reproduce and evolve. Now let us apply these concepts to a process such as the induction of the *lac* operon by lactose in a bacterium, to see how one can naturally and legitimately talk about this in cognitive terms. Thus, for example, we want to see in what sense a bacterium possessing a functional circuit for the controlled catabolism of lactose can be said to be operating on the basis of knowledge.

Evidentally a specific environmental feature that we are concerned with in this case is the presence of lactose in the nutrient medium, together with the absence of glucose. The bacterium first recognizes this by means of the specific receptors on the cell membrane, the lactose permease molecules, some of which are always present due to the noisiness of the control circuit. These transfer lactose from the exterior to the interior of the cell, where it undergoes metabolic conversion by β-galactosidase (also always present at basal levels) to another steric form, allo-lactose, which is the specific inducer. This interacts with repressor, inactivating it and so releasing the *lac* operon from repression, providing cAMP levels are sufficiently high. We have operating here a communication channel between the membrane and the DNA, carrying the initial signal in coded form and resulting in the activation of a specific part of the cellular memory store (DNA). Specific coded information from this store is released (mRNA), which is then translated into proteins. The proteins (permease, acetylase, β-galactosidase) then diffuse to the membranes and are there assembled into structures whose activity results in the transport and catabolic degradation of lactose.

The rules of macromolecular assembly on membranes or other organelles are part of the heuristics for the operational realization of the whole process;

i.e., the system has to 'learn' by trial and error which assembly rules or interactions of proteins with membranes and cytoplasmic elements give the desired result: the transformation of lactose to provide energy and building blocks for biosynthetic processes. This 'learning' process is currently understood in terms of random variations in the DNA memory store, together with variations in the metastable, heritable cellular structures such as membranes and organelles, with a selection for those states which generate greatest biological stability. Heuristic rules in fact operate at all steps in the epigenetic response, since all the recognition events by macromolecules, encoding, transmissions, etc., arise through the historical selection process.

There are no unique solutions to these optimization problems. Thus the rules learned are contingent and to some extent arbitrary. For example, in *E. coli* the three structural genes of the *lac* operon are transcribed simultaneously, whereas in another organism they might be at separate sites and under pleiotropic control, as are the arginine biosynthetic enzymes. The various encodings, transmissions, decodings and interaction rules which make up the entire functional *lac* operon in *E. coli* constitute a useful description of an aspect of the world, which in this case is the fact that lactose is a molecule which can be converted into simpler molecules with a concomitant release of energy. This knowledge is implicit in the causal relationships between the elements which make up the lactose control circuit as described in Figure 2.3 and it becomes manifest when lactose is present and dominant sugars such as glucose are absent from the environment. Thus we may refer to this circuit as a cognitive unit.

Every aspect of the operation of such a unit is in accordance with physical and chemical laws. The reason it is legitimate to refer to such a unit in cognitive terms is that the description of an aspect of the world which is contained in such an entity is not unique: it is only one of many possible descriptions. Thus a yeast cell is also able to catabolize lactose and has a control circuit whose net behaviour is similar to that which operates in *E. coli*. However, the details of the encoding, transmissions, decodings, interaction and assembly rules are different. That which is described is the same, but the description is different. This is a basic characteristic of knowledge, that there are many different ways in which it can be expressed. Thus knowledge about the structure of a house can be described by a set of two-dimensional plans giving various sections through the three-dimensional structure; or it can be expressed by a three-dimensional model which can be taken apart in various ways, and so on. This diversity of possible descriptions of one common set of constraints defining an aspect of the world is a source of immense cultural variety at the anthropological level. In the present context of cellular organization these functional descriptions can be seen as the major source of variety of organismic structure and behaviour, viewed from a cognitive perspective. The very arbitrariness (*a priori*) of the descriptions is one of the difficulties of discovering general laws of biological organization, which must transcend all particular realizations. Organisms by the above definition become knowledge-using systems and as such are sharply distinguished from non-biological ones. The latter systems may have descriptions of aspects of their environment, as a rock fossil contains a description of an aspect of an extinct organism; but this description is of no use to the rock and so it cannot be regarded as a cognitive system. Only organisms, so far as we know, employ useful descriptions of aspects of their world to achieve conditions of stability in relation to that world.

The knowledge expressed in the *lac* operon involves some quite complex coding and decoding operations which have arisen as a result of various functional demands which the species has sought to satisfy in an optimal manner throughout its evolution. A major one is the necessity to stabilize the phylogenetic memory store, which is largely encoded in DNA, and at the same time to have a mechanism for the transmission of this memory to the organism's progeny. The information encoded symbolically in the DNA may be seen in this context as a set of theories or hypotheses about aspects of the world. These hypotheses generate activities (via proteins) in response to specific eliciting signals, and these activities test the validity of the hypotheses (see also a discussion of this viewpoint in Goodwin, 1972). These processes constitute a highly intricate web of molecular interactions within which it is difficult to recognize any simple description which the organism has, for example, about the availability and utility of lactose as a nutrient source. We have to learn to decode the description, just as we do when we are learning a new language.

A biological process which is much easier to read than that expressed in a control circuit of the type described above is the biological clock, discussed in Chapter 4. An organism with a circadian rhythm evidently constains variables with change in an oscillatory manner with a period of about 24 hours. Such a periodicity constitutes, in very obvious terms, a description of the solar cycle. The nature of the description is much clearer in this case than in the previous example because of the direct analogue relationship between the periodicity of the solar day and that of the organismic oscillator, there being very few 'symbolic' or coded elements involved. However, there are once again heuristic rules which define the details of the rhythmic physiological and behavioural activity of any particular organism, establishing the optimal relationship between internal processes and the environmental periodicity. Thus cell division, photosynthetic activity, chloroplast swelling, respiration, phototactic sensitivity, and other cellular properties are optimally timed in relation to the 24-hour environmental cycle and to one another, and provide an easily read statement of dynamic order and harmony within the cell as a result of its evolution in a world with periodicities.

The distinction between knowledge and information becomes clear in terms of the definition given above. Information is always an aspect of knowledge, for the latter invariably involves a selection of a small class of events from a much larger one (e.g. the lactose permease is capable of responding to a small set of sterically similar molecules). What is lacking in the technical definition of information is any reference to its meaning or significance in relation to the behaviour of the system carrying the information. A control circuit, for example, is more than an information channel with encoding and decoding operations. It uses this coded information to carry out useful operations and may thus be said to interpret the information in a particular way. This process of interpretation always involves useful activity in real space-time such as the transport of substances across a membrane, the catalytic transformation of a molecule such as lactose, the synthesis and folding of a protein, etc. The meaning of a signal or a message carrying information is thus revealed in the interpretation process and this takes place within a certain context or environment which involves the action of particular forces. We may thus say that knowledge operates when information is interpreted in a particular space-time context.

This is equally true in symbol-using organisms such as human beings, for whom information can be encoded in symbols whose spatio–temporal structure need bear no obvious relationship whatsoever to elements in the field of action to which they refer. The interpretation process which mediates between the symbol and the field of action defined by its denotation may then employ purely cultural relationships such as rules or conventions, which must be learned. The distinction between a signal and a symbol is then primarily to be found in the fact that the latter has an extra degree of freedom arising from its association with a learning process, whereas the former evokes an instinctive or automatic interpretation, such as the metabolic response to the cAMP 'glucose distress signal' in starving bacteria, discussed in Chapter 2.

Inverse Limit Systems

The functional hierarchy which has been used to describe biological organization is now seen to be describable as a hierarchy of cognitive processes, each functioning on the basis of knowledge about some aspect of the world. Each level of the hierarchy involves behaviour appropriate to events taking place over a particular period of time, this temporal selection resulting from the response times of the different steps in the response. Thus we may say that a particular level of the hierarchy is interested only in processes taking place at a particular rate or within a particular frequency range. We may formalize the relationships between these levels in the following way.

Let us call an event E, and the various functional levels of an organism which can respond to the event, $C_0, C_1, \ldots C_n, \ldots$ These levels of behaviour are related to one another by operations which involve averaging in the sense that C_{n-1} 'sees' and responds to a time-averaged version of what C_n 'sees' and responds to. For example, if the event E is a reduction of the exogenous arginine level in a bacterial culture, then C_{n-1} could be the epigenetic level of the arginine control system, responding to a time-averaged version of what C_n, the metabolic level of the system, responds to. The whole of the arginine control system is a cognitive unit.

We may set up a correspondence between an event and the various levels of response to it in the manner shown below in Figure 7.1. Here the F_ns are

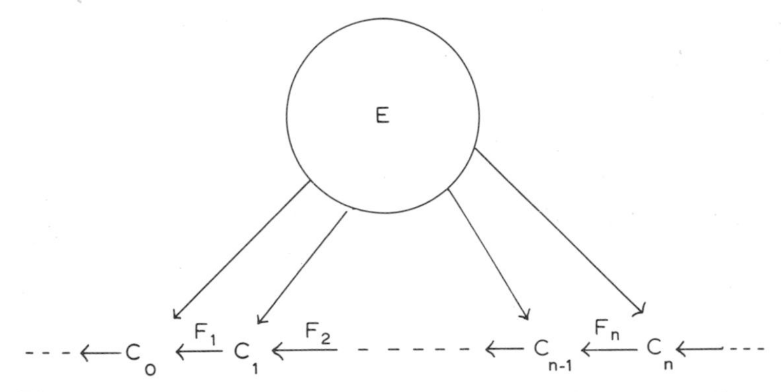

Figure 7.1. A description of the hierarchical organization of organisms in terms of inverse limit systems.

the mappings (involving averaging operations) which relate the C_ns to one another, so that C_0 is the coarsest temporal description of E while C_n is the finest of those explicitly written in the figure. Although I have concentrated upon temporal aspects of hierarchical relationships, such an analysis can be applied to spatial discrimination as well. Then the functions F_n would involve spatial as well as temporal averaging. However, in what follows I will restrict the argument to the time domain. Also, although I have restricted the detailed analysis of hierarchical relationships to very simple organisms such as bacteria, the formalism clearly applies to more complex organisms with many levels of response.

Providing there is a composition rule for the mappings F_n so that C_n maps to C_m $(n > m)$ by the composite mapping

$F_{n,m} = F_m \circ \ldots \circ F_{n-1} \circ F_n$, together with a unit element, then the above sequence is referred to in topology as an inverse limit sequence (Eilenberg and Steenrod, 1952; Hocking and Young, 1961). To make things much clearer let us consider an example in which the T_n are pure time averaging operators. Let $X(t)$ be a time-dependent variable which could be, for example, the concentration of some nutrient in a bacterial culture, or the intensity of light at the surface of a pond. Let the time at which the variable acts as an input to some averaging operator (labelled i) in an organism be τ_i, while the time at which there is a response from the operator is t_i. Write

$$s_i = t_i - \tau_i$$

so that we have

$$X(\tau_i) = X(t_i - s_i).$$

Now let us define the time average of X as carried out by this operator to be

$$\langle X \rangle_i(t_i) \equiv \int_0^\infty \omega_i(s_i) X(t_i - s_i) \, ds_i. \tag{7.1}$$

This is the same type of equation as the renewal equation of Chapter 4. Here the function $\omega_i(s_i)$ characterizes the nature of the averaging operation carried out by the unit under consideration, such as a control circuit. In order to satisfy the requirements of an averaging process, we must have

$$\omega_i(s_i) \geqslant 0 \text{ for } s_i \geqslant 0$$

and

$$\int_0^\infty \omega_i(s_i) \, ds_i = 1$$

This definition is taken from Kornacker's (1972) interesting discussion of measuring processes in living systems and their use in the formulation of a generalized statistical mechanics, with the trivial difference that the lower bound of the integration is here taken as 0 instead of ∞, since $s_i > 0$.

Suppose that two different averaging operators, i and j, act in sequence on $X(t)$. Then we get

$$\langle\langle X \rangle_i\rangle_j(t_j) = \int_0^\infty \omega_j(s_j) \langle X \rangle_i(t_j - s_j) \, ds_j$$

But

$$\langle X \rangle_i(t_j - s_j) = \int_0^\infty \omega_i(s_i) X(t_j - s_j - s_i) \, ds_i$$

and therefore we get

$$\langle\langle X\rangle_{\mathrm{i}}\rangle_{\mathrm{j}}(t_{\mathrm{j}})=\int_0^\infty \omega_{\mathrm{j}}(s_{\mathrm{j}})\,\mathrm{d}s_{\mathrm{j}}\int_0^\infty \omega_{\mathrm{i}}(s_{\mathrm{i}})X(t_{\mathrm{j}}-s_{\mathrm{j}}-s_{\mathrm{i}})\,\mathrm{d}s_{\mathrm{i}}.$$

This is called a double convolution integral. Generalizing the result to the case of a sequence of averaging devices, we find that

$$\langle\ldots\langle\langle X\rangle_m\rangle_{m+1}\ldots\rangle_n(t_n)=\int_0^\infty \omega_n(s_n)\,\mathrm{d}s_n\int_0^\infty \omega_{n+1}(s_{n+1})\,\mathrm{d}s_{n+1}\ldots$$

$$\ldots\int_0^\infty \omega_m(s_m)X(t_n-s_n-s_{n+1}\ldots-s_m)\,\mathrm{d}s_m \qquad (n>m)$$

thus obtaining a particular example of a composition rule for successive operators in the scheme represented by Figure 7.1.

It is instructive to see how our analysis of the dynamic behaviour of molecular control circuits carried out in previous chapters fits into the general descriptive scheme in terms of averaging operations. Let us consider a metabolic control circuit of the type described in Chapter 1, involving end-product inhibition with stoichiometry n. Let this end-product, an amino acid, say, be present in the nutrient medium in concentration $z_{\mathrm{i}}(t)$. This will be undergoing fluctuations with some characteristic spectral properties (i.e. some distribution of amplitudes and frequencies of the fluctuations), so $z_{\mathrm{i}}(t)$ is a stochastic (random) variable. This amino acid will be transported into the cell and will contribute to the intracellular pool of end-product which will be labelled y_{i}, so the fluctuations in z_{i} will be transmitted to the pool. The objective of the control circuit is to maintain y_{i} at a relatively constant value and so to counteract the fluctuations coming from $z_{\mathrm{i}}(t)$. This is precisely what is achieved by the negative feed-back control loop, providing the relaxation time of the loop matches the frequency of the fluctuations in z_{i}. Let us see how this comes about.

Consider the simplest case of a one-step negative feed-back circuit in which the product of an enzyme inhibits catalytic activity of the enzyme. The dynamics of the control circuit may then be written in the form

$$\frac{\mathrm{d}y_{\mathrm{i}}}{\mathrm{d}t}=\frac{a_{\mathrm{i}}}{b_{\mathrm{i}}+y_{\mathrm{i}}^n}+c_{\mathrm{i}}z_{\mathrm{i}}-d_{\mathrm{i}}y_{\mathrm{i}} \qquad (7.2)$$

where the Hill equation which has been used for the control term, z_{i} is transported into the cell at the rate $c_{\mathrm{i}}z_{\mathrm{i}}$, and y_{i} is the intracellular concentration of the amino acid, which is being removed for biosynthetic purposes at the rate $d_{\mathrm{i}}y_{\mathrm{i}}$. Consider the behaviour of this control circuit in the neighbourhood of the steady state, $\bar{y}_{\mathrm{i}}$, using the usual assumption that variations from this state are small. Then the equation can be linearized in terms of the small quantity $x_i = y_i - \bar{y}_i$ to give

$$\frac{dx_i}{dt} = c_i z_i - \gamma_i x_i \tag{7.3}$$

where $\gamma_i = \alpha_i + d_i$ and α_i is the slope of the control function $a_i/(b_i + y_i^n)$ at the steady state, $\overline{y_i}$. As shown in Chapter 1, an important factor in determining the size of this quantity is the stoichiometry of the inhibition, n.

Equation (7.3) can now be solved in the following manner. Rewriting the equation in the form

$$\frac{dx_i}{dt} + \gamma_i x_i = c_i z_i$$

and observing that

$$\frac{d}{dt}(x_i e^{\gamma_i t}) = e^{\gamma_i t}\left(\frac{dx_i}{dt} + \gamma_i x_i\right),$$

the equation can be rewritten as

$$\frac{d}{dt}(x_i e^{\gamma_i t}) = c_i e^{\gamma_i t} z_i(t).$$

Integrating both sides, we find

$$x_i e^{\gamma_i t} = \int_0^t c_i e^{\gamma_i \tau} + C,$$

where C is the constant of integration. This constant can be written

$$C = \int_{+\infty}^{0} c_i e^{\gamma_i \tau} z_i(\tau)\, dt,$$

i.e., the 'history' of the process up to time zero, so that the solution takes the form

$$x_i(t) = e^{-\gamma_i t} \int_{-\infty}^{t} c_i e^{\gamma_i \tau} z_i(\tau)\, d\tau$$

$$= c_i \int_{-\infty}^{t} e^{-\gamma_i (t-\tau)} z_i(\tau)\, d\tau$$

Now write $t - \tau = s$ to give

$$x_i(t) = c_i \int_0^{\infty} e^{-\gamma_i s} z_i(t-s)\, ds. \tag{7.4}$$

It is evident that this has the same form as (7.1) if we define

$$\omega_i(s_i) = Ac_i e^{-\gamma_i s},$$

where A is a normalising factor which is required to satisfy the constraint

$$\int_0^\infty \omega_i(s_i)\, ds_i = 1.$$

This gives us

$$A\int_0^\infty c_i e^{-\gamma_i s}\, ds = 1, \qquad \text{whence} \quad \frac{Ac_i}{\gamma_i} = 1 \quad \text{and so} \quad A = \frac{\gamma_i}{c_i}.$$

This result shows that for a linear ordinary differential equation of type (7.3) the averaging operator is

$$\omega_i(s) = \gamma_i e^{-\gamma_i s}, \qquad \text{with } \gamma_i = \alpha_i + d_i.$$

Evidently the properties of this function are determined by α_i and d_i, whose values are specified by the response characteristics of the control circuit and the mass action effect, respectively. For a metabolic process of the type described by equation (7.2), γ_i can be estimated to be about 10^{-2}, as expected from the considerations of Chapter 1. A control circuit with a particular relaxation time will clearly smooth out fluctuations in $z_i(t)$ which fall within its frequency range. We may say, resorting to cognitive language, that a control circuit with a certain relaxation time expects there to be environmental fluctuations with a certain range of frequency. A measure of cognitive performance in the circuit is then given by its success in quenching these fluctuations. The control circuit has evolved to function with a certain expectation and the success of its operation can be measured as the extent to which this expectation matches the behaviour of $z_i(t)$. Although the analysis has been presented in linear form only, the general principles underlying the use of the operators $\omega_i(s_i)$ in equation (7.1) should now be clear.

The formal range of integration in equation (7.4) is $(0, \infty)$, but there will be an effective averaging time which is determined by γ_i and hence by the characteristic response time of the control system, beyond which no significant averaging will be effected. Call this time T_m for the metabolic system, so that we can write

$$x_i(t) = c_i \int_0^{T_m} e^{-\gamma_i s} z_i(t-s)\, ds = c_i \langle z_i \rangle_{T_m}(t) \tag{7.5}$$

The parameter c_i includes the concentration of any permease bringing the exogenous metabolite into the cell, while γ_i includes enzyme concentrations. These may be taken as constants over the time interval T_m. However, $x_i(t)$

now becomes the input to the epigenetic system and the variations in $c_i(t)$ and $\gamma_i(t)$ must be taken into account over longer time periods than T_m. This gives rise to cross-correlation terms, $\langle c_i(t)\langle z_i\rangle_{T_m}(t)\rangle_{T_e}$ in the time averages of epigenetic processes, where T_e refers to the relaxation time of the epigenetic system. It is from such cross-correlation expressions that one obtains evidence of coherent dynamical behaviour reflecting the capacity of the system to resonate with fluctuations and so to operate in a highly efficient manner. This property of hierarchically-organized systems has been repeatedly emphasized by Kornacker (1969*a*, 1969*b*, 1972), who sees it as the major source of dynamical order in biological systems and actually defines cognitive processes in terms of the calculation of correlations. McClare (1971, 1972) has presented a convincing argument for the view that dynamical correlations and resonance are the source of the extra-ordinarily high efficiency of energy transfer in biological systems, and has suggested an approach to biological energetics via resonant mode interactions.

While lower levels of the hierarchy generate inputs to higher levels, the latter define dynamical constraints or parameters for the former. This is evident from equation (7.5), where $c_i(t)$ is acting as a parameter for the metabolic variable $x_i(t)$. At the epigenetic level, not only does one get cross-correlations between $c_i(t)$ and $\langle z_i\rangle_{T_m}(t)$, one must also consider correlations with environmental inputs coming directly to this level, as indicated by the arrow from E to C_{n-1} in Figure 7.1. An example of such an influence is a daily light-dark cycle which may affect gene activities directly via a change in temperature, hydrogen ion concentration, etc. Dynamical couplings in a complex system then include correlations between inputs from lower levels and those from the environment with the internal dynamics of the level itself. These relationships are thus correlative rather than causal, each level and each component within a level exhibiting a degree of autonomy. One can evidently get much more complex and interesting behaviour from these interactions and correlations than the simple quenching of environmental fluctuations achieved by a matching of relaxation times to fluctuation frequency, as described above in relation to simple control circuit operation.

Applying these ideas to an organism with circadian organization, we may say, cognitively speaking, that the system expects a near 24-hour rhythm in one or more environmental variables and a measure of its cognitive success is given by the cross-correlation or coherence function of the system variables and the environmental periodicity. Again, in considering the organization of events during the cell cycle, the same criterion of cognitive performance may be used, involving a measure of the stability of the frequency and phase relationship between constituent activities such as initiation of DNA replication, budding (in yeast, say), nuclear migration and cell division. The budding process, which can occur independently of DNA replication in certain mutants (Hartwell *et al.*, 1974), has a certain expectation in relation to chromosome replication, mitosis and nuclear migration, and conversely. The degree of cognition manifested by the whole is given by the cross-correlation between them, which measures this coherence.

The situation becomes more complex when spatial as well as temporal factors are considered, but the same principles apply. The recognition of molecules by proteins is an obvious example of a response to a signal with a spatial energy distribution and the cognitive success of this recognition depends upon a complex resonance matching between spatio-temporal frequencies in the protein and in the molecule with which it interacts. The resonance matching between a protein and a ligand is possible because of certain invariants of molecular structure and activity, and these features are used to initiate specific state changes in the protein-ligand complex which contribute to the overall stability of the organism. The protein may thus be said to have a useful description of the molecule with which it interacts.

It is thus evident that the cognitive view of organismic behaviour uses a decomposition of the system into parts each of which operates in terms of a description of a certain aspect of itself and/or its environment. The most obvious example of a molecule with a useful description of itself is the DNA double helix. The parts of the cognitive system have a certain autonomy, but the whole is reconstituted by considering the couplings or interactions between variables, seen as signals, which result in co-ordinated behaviour measured by cross-correlations. The population of partially-autonomous observers and operators thus becomes a whole.

As stated above, the operators F_n will in general be more complex than simple averaging operators of the type described by equation (7.1). We may keep the picture quite general by taking the descriptions, C_n, to be categories and the F_n to be functors (McLane, 1971). Then each category constitutes a particular description of the event, E, and an organism uses some finite set of these descriptions, $\{C_n\}$. Extending the sequence in Figure 7.1 indefinitely in both directions gives a formal scheme for the potential set of descriptions of any event and hence the set from which organisms may make their selection according to their evolutionary histories. We see that no finite set of C_n constitutes complete knowledge about E, and that only the space defined by the set of all sequences $(\ldots c_{n-1}, c_n, c_{n+1} \ldots)$, where c_n is a point of the space C_n, called the product space and written

$$\mathbb{P}_{-\infty}^{\infty} C_n,$$

can be said to contain a complete description of E. This is referred to as the inverse limit space, and it is with the properties of this space that the topology of inverse limit systems is concerned. In the particular example of the averaging operators considered above, the mappings T_n are the functionals, equation (7.1), which map functions of time, $X(t)$, onto other functions of time. These latter functions make up the spaces of Lebesgue integrable functions, which are infinite dimensional Banach spaces.

The reason for using inverse limit systems to formalize the relationships existing between levels of the functional hierarchy used in our analysis of biological organization is to clarify two points. The first is to show that in cognitive terms any organism functions on the basis of a limited set of descriptions of the world, selecting some subset from the inverse limit space which can be said to contain all possible descriptions. The generality of the scheme obtained when one uses categories for the C_n and functors for the F_n is very high indeed, and is unlikely to be of value in the consideration of physiological organization where the primary constraints operating in ordering the hierarchy are the spatial and temporal ones considered above. However, when one comes to the domains of behaviour and psychology, particularly the latter where symbolic descriptions of the world come into operation and the relationships between a description and the domain of action referred to become more and more subtle, then the level of abstraction required to identify the nature of the correspondences or the mappings between levels is likely to require this high order of mathematical structure. The potential power of the inverse limit system approach will then be required to avoid trivializing the analysis of the psychological domain by a facile reduction to physiological principles, while still preserving the principle of basic structural and functional correspondences between them. At this level of description the system represented by Figure 7.1 becomes a lattice of interactions of arbitrarily high dimensionality and functional ordering.

The second reason for using a general topological model such as the inverse limit system to represent organismic hierarchies is to exploit the mathematical structure of this theory in relation to a study of certain basic properties of cognitive processes as defined here. I am far from competent to pursue the latter problem, but there is at least one interesting aspect of topology which seems relevant to the cognitive study of organisms. This insight, as well as the suggestion that inverse limit systems be used in the manner described above, comes from Dr Peter Antonelli.*

There is an area of topology called Čech homology theory whereby, roughly speaking, one may construct minimal covers of spaces or categories, C_n. These minimal covers give the simplest characterization of the topological system under consideration, so that we may say that the essential

* Department of Mathematics, University of Alberta, Edmonton, Alberta, Canada.

structure of a description can be extracted by the Čech homology procedure. This constitutes therefore a minimization operation over descriptions which involves the selection of that which is simplest or, we might say, most elegant within given constraints. In a cognitive context, such a procedure is of profound interest, for by its means we have a rigorous approach to the problem of characterizing a description in terms of maximal simplicity or elegance. Whether or not this is a biologically appropriate procedure requires much deeper study, but it seems a promising direction in which to pursue questions relating to creativity and intuition, which are often regarded as beyond exact definition.

Co-operative Interactions and Phase Transitions in Cognitive Systems

Any level of organismic behaviour, represented in Figure 7.1 as a category, C_n, will in general be a very complex space of functions. However, a prominent feature of organisms is that what emerges from this complexity is relatively simple, ordered behaviour. One way of understanding this emergence of simple coherence from multiple interactions at a particular level of organization is in terms of co-operative processes or collective phenomena. We have already considered a very simple case of co-operative behaviour in relation to the interactions of subunits in a multimeric protein in Chapter 1. Now I shall describe a somewhat more complex and interesting example of a co-operative process which illustrates how interactions between units can give rise to sharp discontinuities of state known as phase transitions. This gives a natural introduction to a more general study of such discontinuities, known as catastrophe theory, which has been developed by the mathematician René Thom and applied to biological processes, particularly to morphogenesis. By embedding the theory in the general descriptive framework of inverse limit systems and cognitive processes, a fairly comprehensive view of the organismic process will emerge. The goal of this development will be an understanding of generative processes, those in which spatially and temporally ordered systems emerge in an orderly sequence from much simpler ones which contain the potential for such an emergence, as the tadpole from the egg. This potential exists in the useful descriptions of self and of environment which the egg contains, descriptions which become manifest in space and time as development proceeds.

The process which I will now analyse as a phase transition is the development of excitability in biological membranes. To make the example specific and to relate it to material already discussed in this book, I will consider the development of metabolic excitability in a membrane of the type considered in the wave model. This problem is also treated in a paper by Goodwin

(1974). Referring to Figure 5.4, it is evident that if the density of enzymes, E_1, is too low, no activity wave could propagate because the activity of one enzyme would then be insufficient to excite its neighbours. The exact relationships between enzymes will depend on the stoichiometry of the metabolic reaction producing V from U, on the effect of V on the equilibrium between bound and free ions, on the concentration of the ions, on the temperature, etc. The density of enzymes in the membrane will be dependent upon the balance between the rate of incorporation into membrane structure and rate of loss and dilution (if new membrane is being formed). We can represent all these factors by means of a chemical potential, μ, so that the density of enzyme is determined by this quality.

Let us suppose that there are sites for enzyme E_1 in the membrane, and that their mean number per unit area is constant. We know very little about the nature and distribution of such sites, whether they have a geometrical order determined by other structures in the membrane, or whether they are distributed at random. It seems probable that the real situation is intermediate between these two: that there are constraints on the distribution of potential enzyme sites, but that these do not generate a rigid geometry. Such considerations do not affect the qualitative model which will be developed, but they do have an effect upon the geometry of wave propagation on the excitable membrane; for example, whether the waves tend to propagate in concentric circles from a point of excitation, or whether in spirals. These alternative consequences will not be considered here.

We use the binary variable ε_i to designate the state of site i. This variable takes the value 0 or 1 according as the site is unoccupied or occupied, respectively. The model is based upon what is known in physics as the lattice gas description of condensation, which is in turn based upon the Ising model of ferromagnetism, using the mean field approximation. This assumes that each unit in the system under consideration (in our case, an enzyme) experiences a mean field of force due to the presence of other units, and the essence of the procedure is to find an equation which expresses this field in terms of itself (the self-consistent field equation). The field we are considering arises from interactions assumed to occur between enzymes occupying membrane sites and those entering the membrane, the presence of an enzyme at a site facilitating the entry of an enzyme into the membrane at a neighbouring site. We designate this as $\rho = \langle \varepsilon_i \rangle$, the mean value of the occupancy averaged over all the sites. Let the total number of occupied sites (i.e. enzymes) in the membrane be N, so that

$$\sum_{i=1}^{s} \varepsilon_i = N,$$

the sum being over all sites. And finally, let the interaction between two enzymes, E_i and E_j, be u_{ij}.

The appropriate function to use in determining average values for such a system is known as the grand partition function, which allows for variations in both N, the total number of active enzymes on the membrane, and in the energy of the system, deriving from interactions. At this point let me refer the reader to books dealing with the Ising model and the lattice gas (e.g., R. Brout, 1965; K. Huang, 1963), where it is shown that the energy function (Hamiltonian) for this system has the form

$$H - \tfrac{1}{2}\sum u_{ij}\varepsilon_i\varepsilon_j - N\mu. \tag{7.6}$$

The self-consistent equation for ρ which is obtained from this expression via the grand partition function is

$$(2\rho - 1) = \tanh\,[\beta/_2(\rho u(0) + \mu)] \tag{7.7}$$

where $u(0)$ is the Fourier transform of $\sum u_{ij}$ taken at the point O. The assumption here is that the interaction field is everywhere the same, $u_{ij}(R)$ being translation invariant. The parameter β is $1/kT$, k being Planck's constant and T physical temperature. These enter the equation as a result of the assumption that the mean energy is constant, hence that there is a defined temperature. It is this assumption that implies that we are here dealing with a system that is, with respect to the processes under consideration, at equilibrium. Thus although we know that membranes are maintained by an active energy flow in living systems, we are ignoring this and treating the processes of enzyme association and dissociation from membranes and interactions between enzymes, as if they were simple chemical reactions describable by a chemical potential and constant interaction terms. Thus we are looking at the whole process in terms of chemical equilibrium. Providing the fluxes in the system are sufficiently small, this is legitimate.

Now let us look at the implications of this model. What we are interested in is the possibility of a rather sharp transition from a condition of non-excitability to one of excitability for particular values of the temperature, T, and the chemical potential, μ, which determines the density of enzymes on the membrane. An example of this type of transition in a developing system may be provided by regenerating *Acetabularia* plants. In their electrophysiological studies, Novak and Bentrup (1972) observed that some 8–10 hours after the onset of development, spontaneous action potentials appeared and persisted throughout the period of regeneration. This is a manifestation of some kind of excitability in the system, although it may not be identical with that described here.

The region of the relation equation (7.7) that is of interest with respect to a state transition of this type is the part known as the coexistence region. In gases this is where a transition from gas to liquid can occur and in ferromagnets it is where spontaneous magnetization can appear. The value of the potential making this possible in our case is $\mu = -u(0)/2$. Since μ is defined relative to an arbitrary constant, the negative sign here has no significance. This value substituted into equation (7.7) gives us the equation

$$(2\rho - 1) = \tanh[1/4\beta u(0)(2\rho - 1)]. \tag{7.8}$$

Now what we find is that if β is sufficiently large (T sufficiently small), this equation has three roots; whereas for small β it has a single root which is the trivial solution, $2\rho - 1 = 0$, or $\rho = \frac{1}{2}$. This is made clear by Figure 7.2, where the line $y = 2\rho - 1$ and the curve $y = \tanh[\frac{1}{4}\beta u(0)(2\rho - 1)]$ are plotted on the same axes. Curve (1) has a value of β which gives a slope to the hyperbolic tangent at $\rho = \frac{1}{2}$ which is greater than 2, while in curve (2) the opposite is true.

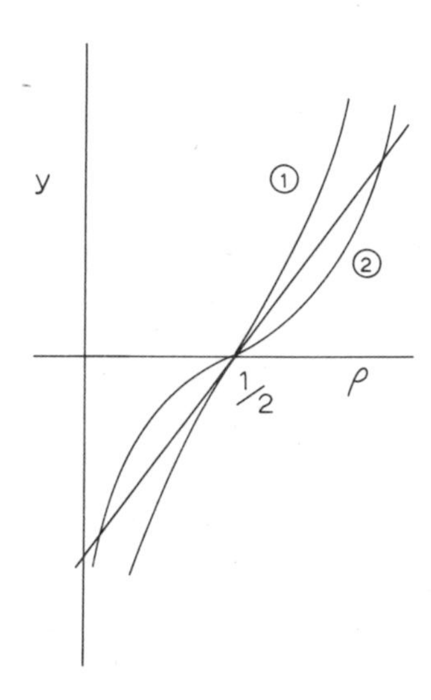

Figure 7.2. Graphs of the functions $y = 2\rho - 1$ and $y = \tanh[\frac{1}{4}\beta u(0)(2\rho - 1)]$ showing the way in which variations in β cause the appearance of multiple solutions in ρ to equation (7.8).

The critical value of β (call it β_c, defining T_c) is where the slope at this point is 2.

If now we plot ρ as a function of T, we get the curve shown in Figure 7.3. Above the critical temperature, T_c, the only root is $\rho = \frac{1}{2}$, which is an indeterminate condition of the membrane: neither excitable nor non-excitable. But below T_c, the membrane can change from a state of excitability to one of non-excitability depending upon ρ, the mean interaction field. Thus if the system starts at P in the non-excitable region and ρ increases, (more enzyme molecules are added to the membrane) then the system will find itself at the point P' on the coexistence curve. This is an unstable point, and the system will rapidly undergo a transition to Q', thence entering the region of excitability. As the temperature is decreased, the transition from non-excitability to excitability takes place at smaller and smaller values of ρ, the mean excitation field; i.e., thermal noise interferes less and less with the processes resulting in excitation.

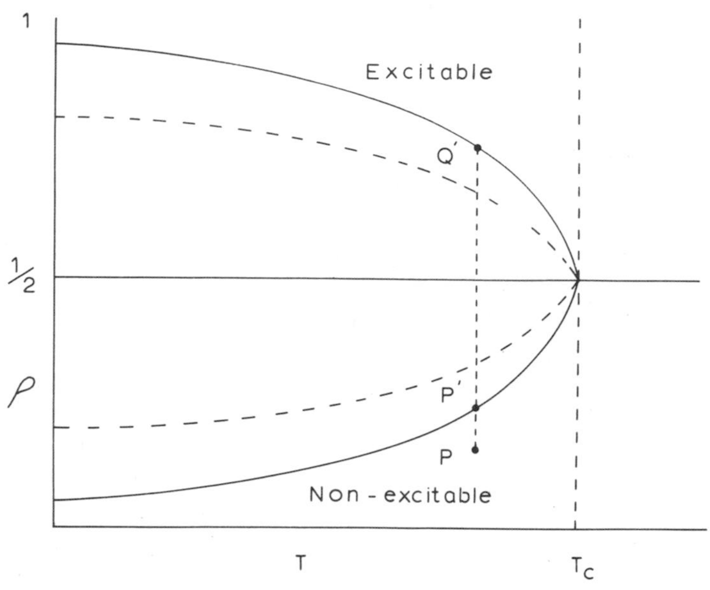

Figure 7.3. The coexistence curve for non-excitable and excitable states in a membrane below the critical temperature, T_c, according to the model described in the text.

Now clearly in a biological system there will also be a lower temperature bound to the domain of excitability as well. Thus the coexistence curve should close in on itself again, providing the process of loss of excitability is describable also as a phase transition. This could be modelled by making further assumptions about, say, the temperature-dependence of the quantities $u(0)$ and μ at lower values of T. However, this does not seem worthwhile in view of the paucity of experimental data on such transitions in biological systems and the possibility that quite different factors from those considered here enter into the process. We do not really know if the spontaneous appearance of action potentials in *Acetabularia* is correctly described as a phase transition, since the temperature-dependence of the process has not, to my knowledge, been studied. This would be of considerable interest. However, the main purpose of introducing the model is to illustrate how discontinuous changes of state can occur in systems consisting of many interacting units, and how they can be explored analytically. The mean field treatment given above provides a particularly simple procedure for describing how co-operative interactions, in this case the assembly of protein molecules in membranes, can result in sudden changes of behaviour which are of biological importance. A basic proposition arising from this is that many of the most significant changes of state occurring in biological systems are of this general type: that important aspects of morphogenesis and behaviour can be described by such discontinuities, which occur in a highly controlled and orderly manner. These, then, arise within the context of dynamic change in an overall stable, regulated system, stabilized by the more continuous homeostatic mechanisms discussed earlier in this book. A discontinuous change of state occurring within a particular level of the hierarchy, C_n, will result in the generation of new collective variables at that level, so that the 'signals' transmitted via the function F_n will be altered. In the above example, before the development of excitability in the membrane, no metabolic waves can travel and hence, in terms of the wave model of Chapter 5, no gradient can be formed. Thus no space-ordering signal can be generated until this phase transition occurs and spatial differentiation can arise only from that which is present initially in the system. This may be considerable, but it will be of mosaic type and unable to regulate. Once membranes become excitable, however, a new type of behaviour arises in the metabolic system, namely metabolic wave propagation and regulating, spatial-ordering information in the form of membrane-bound gradients can control epigenetic state changes. Thus we see in what ways phase transitions can form an important feature of the generative process we call development and how they can enter into the descriptive context of Figure 7.1 as discontinuous state changes within particular levels or the hierarchy, simplifying the dynamics as a result of co-operative interactions between units.

Phase Transitions and Catastrophe Theory

The model of the transition from non-excitability to excitability in membranes described above is essentially the same as that used to describe the condensation of a gas; and it is the simplest 'microscopic' treatment of this process. There is a more phenomenological or macroscopic approach to this problem proceeding from the van der Waals modification of the ideal gas law, which introduces a small correction term to take account of the incompressibility of the gas molecules. This equation has the form

$$\left(p+\frac{a}{V^2}\right)(V-b)-NkT=0. \tag{7.9}$$

For different values of T, the p *versus* V curves are of the form shown in

Figure 7.4. There is a particular value of T, called the critical temperature, T_c, giving a point of inflexion with zero slope in the curve. Above this temperature, there is a single-valued relationship between p and V, but

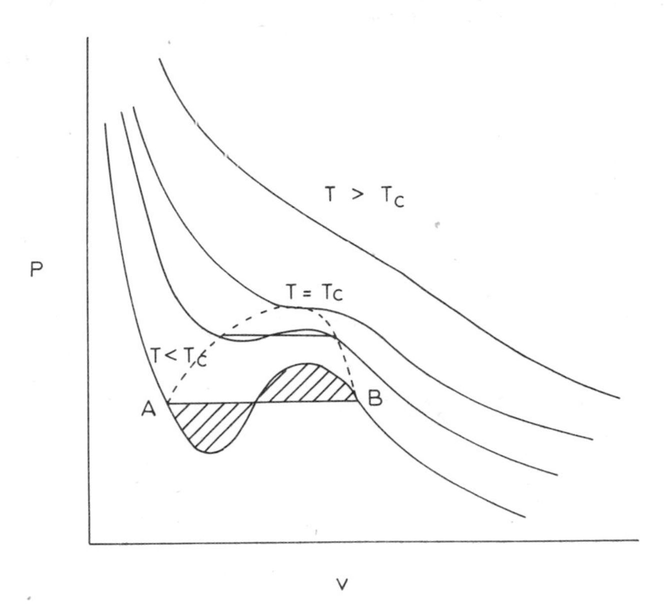

Figure 7.4. Pressure-volume curves of a gas in the region of the critical temperature, showing the use of Maxwell's convention for interpreting the condensation region.

below it there are three roots in V for any given p. This is the condensation region, where the gas suddenly undergoes transition to a liquid with a rapid change of volume, at constant pressure and temperature. The convention used to define where on the cubic this transition takes place is known as Maxwell's convention or rule. The procedure is to draw a line through the cubic such that the areas cut off above and below the line are equal, such as the line AB, the shaded areas being equal. This line is taken as the pressure at which condensation occurs for a given curve (hence given temperature). The curve joining the points where such lines meet the p, V curves is the coexistence curve, shown as a dotted curve in Figure 7.4, which has the same shape as that in Figure 7.3. It is somewhat easier to see the relationship between these curves if one considers the three-dimensional p, V, T surface shown in Figure 7.5. This surface has a fold in it which starts at the critical temperature, T_c. The coexistence curve is shown dotted, and the projection of this curve onto the T, V plane clearly gives a curve like that shown in Figure 7.4. In the lattice gas description of condensation, the variable ρ is the density, which is proportional to V^{-1}, so the shape of the curve remains essentially the same as the T-V projection. This topic is treated in any text-book dealing with phase transitions, but a very complete one is by Stanley (1971).

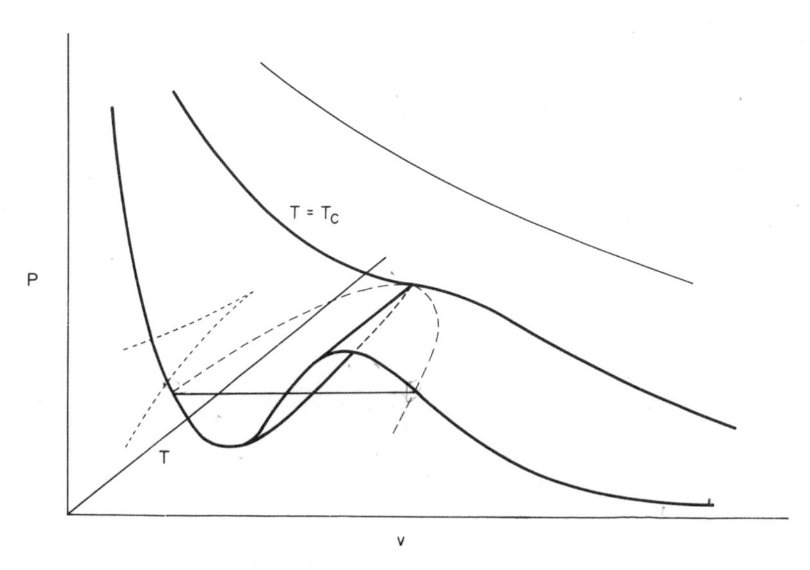

Figure 7.5. The pressure–volume–temperature surface of a gas in the region of the critical temperature, showing the coexistence region and the shape of the domain of multiple roots.

Now the surface shown in Figure 7.5 belongs to a family of surfaces which can be described by the equation

$$x^3 + u_1 x + u_2 = 0 \tag{7.10}$$

in which u_1 and u_2 are parameters which can take different values. The van der Waals equation, written in terms of variables whose origin is the critical point on the surface (p_c, T_c, V_c), is

$$x^3 + \tfrac{1}{3}(8t + y)x + \tfrac{1}{3}(8t - 2y) = 0$$

where $y = p - p_c$, $x = 1/V - 1/V_c$, $t = T - T_c$, as has been shown by Fowler (1972) in an interesting treatment of the van der Waals equation in relation to catastrophe theory. Thus in this case $u_1 = \frac{1}{3}(8t + y)$ and $u_2 = \frac{1}{3}(8t - 2y)$.

If equation (7.10) is integrated with respect to the variable x, we get the expression

$$V(x) = \frac{x^4}{4} + \frac{u_1 x^2}{2} + u_2 x \tag{7.11}$$

taking the constant of integration to be zero. This is called by Thom (1970*b*) the universal unfolding of the cusp or the Riemann–Hugoniot catastrophe, whose germ or organizing centre is $x^4/4$. There are seven elementary catastrophes in Thom's theory, each of which is defined in terms of a potential function such as V in equation (7.11) from which is obtained a particular surface. A basic proposition in this theory is that certain biological processes may be described in terms of the operation of forces drawing the system towards the minimum values of the potential surface, as is frequently done in physics. The surface shown in Figure 7.5 describes that on which a gas 'moves'; i.e., the state of the gas is found, at equilibrium, on the surface. As one variable changes, so the others change according to equation (7.10), obtained from equation (7.11) by differentiation with respect to x and setting this derivative equal to zero to define the extrema:

$$\frac{dV}{dx} = x^3 + u_1 x + u_2 = 0,$$

thus returning to equation (7.10). Thus Thom sees a dynamical process such as the behaviour of a gas under various conditions of pressure, volume and temperature as describable by a surface derived from the universal unfolding of an ordinary catastrophe. So far, this is no more than an embedding of familiar physics into a new mathematical structure. We have seen that there are certain parts of these state surfaces which are of particular interest, such as the regions defining coexistence domains for critical phenomena such as phase transitions. These are the regions of multiple roots (triple in this case). Such regions are defined mathematically by the extrema of the surface shown in Figure 7.5, multiple roots in V occurring whenever p and T lie within the cusp-shaped domain obtained by projecting the extrema of the surface onto the p, T plane, as shown in the figure. This is the region within which phase transitions of shock waves, as Thom calls them, can occur. Such shock waves he regards as the essential basis of the generation of form, of morphogenesis, just as a gas–liquid phase transition describes a change in the morphology of a gas.

One of the problems with this theory is the question of conventions: how is one to interpret the behaviour of the system in the neighbourhood of a mathematical singular point? We have seen that one such convention is that ascribed to Maxwell, but there are others, as discussed by Fowler (1972). Thom (1970*a*) also discusses Maxwell's convention, its defects and the possibility of using other rules of physical interpretation of mathematical singularities, as well as the question of fluctuations and how to represent them in terms of catastrophe theory. A very interesting paper on the purely dynamical interpretation of catastrophe theory is that by Zeeman (1972).

However, the most glaring difficulty in the application of catastrophe theory to morphogenetic processes seems to me to be the problem of interpreting, not the dynamical aspects of the theory, but its spatial implications. Thom (1970*b*) gives spatial as well as temporal interpretations of the elementary catastrophes which derive from the premise that spatial variables enter into the parameters describing the unfolding of a catastrophe. No conventions are given for this, so that there appears to be complete freedom of choice, despite spatial interpretations which imply particular constraints on the way in which they enter the functions. Thus, for example, Thom (1970*b*) applies the swallow-tail catastrophe to the process of gastrulation in amphibia, whereas it could equally well be argued that the cusp catastrophe, unfolding within a particular spatial geometry, is quite sufficient to describe this process. In fact, Zeeman (1974) has now done just this.

I think that the only way to proceed with this problem is to work with the catastrophes in relation to particular biological processes and to see if they do provide useful descriptions (i.e., to see if they help our own cognitive activities). The appropriate conventions for their interpretation may then emerge from the context in which they are being used. One unfortunate aspect of catastrophe theory is its name, which reflects a particular non-biological perspective in dynamical discontinuities. The implication is that these give rise to disorder, rather than accompanying an increase in order. In an engineering context the term seems singularly appropriate, as Thompson's (1975) very interesting studies show; but in biology the discontinuities of state underlying morphogenesis are not in any sense catastrophic. As long as biologists must import theories arising from non-biological contexts, we must suffer these terminological anomalies. The same remarks apply to the theory of dissipative structure developed by Glandsdorff and Prigogine (1971). This term arises from a thermodynamic perspective which takes the closed system as the reference and an open, far-from-equilibrium system to need the description 'dissipative'. All real systems are in fact open, closure being a convenient fiction and biological systems are not only open, but ordering. Energy dissipation is an axiomatic feature of an open system, but it gives no insight into the origins of biological order. In moving towards a theory of generative process in cognitive systems, I hope it may be possible to define a more appropriate context for the study of biological organization and dispel the misleading terminology of catastrophe theory and dissipative structures.

Collective Modes in Rhythmic Processes

I shall now consider ways of describing certain forms of temporal order which have been discussed in previous chapters by means of potential functions belonging to the families of elementary catastrophes. Let us take first the case of the continuous culture of bacteria growing in a chemostat and synchronized by means of periodic phosphate addition, as discussed in Chapter 4. Synchrony in such a culture is controlled by the strength of the periodic stimulus, as shown by Goodwin (1969*c*). The strength or intensity will be represented by the parameter λ. The basic feature of the process is that at a particular value of this parameter, synchrony of cell division begins to appear in the system and this grows with increase in λ. The frequency of the synchronous variation in cell number is determined by the growth conditions in the chemostat, the genetic characteristics of the cells and the frequency of the periodic forcing function. To represent this appearance of a population periodicity, we may use a potential function of a single variable which measures the collective mode amplitude in the culture; i.e., the mean amplitude of the forced oscillation. Such a function is

$$\phi(r) = -a(\lambda - \lambda_c)r^2 + br^4 \tag{7.12}$$

where r is the mode amplitude, λ_c is the critical value of the forcing parameter, and a and b are positive constants. This potential function has been used to describe the appearance of a collective mode in a laser in response to an external energy source, the laser pump (Graham, 1973),* a process quite analogous to that under consideration. From the potential function equation (7.12) we may obtain the dynamics governing r, the amplitude of the synchronous oscillation, by the usual procedure for obtaining differential equations from potential functions, viz.

$$\frac{\mathrm{d}r}{\mathrm{d}t} = -\frac{\partial \phi}{\partial r} = 2a(\lambda - \lambda_c)r - 4br^3 = 2r[a(\lambda - \lambda_c) - 2br^2].$$

Evidently if $\lambda \leqslant \lambda_c$, the stable steady state solution is $r = 0$, in which case there is no macroscopic or population amplitude; but if $\lambda > \lambda_c$, the stable solution is

$$r = \sqrt{\frac{(a(\lambda - \lambda_c)}{2b}}.$$

We need two variables to define an oscillation, and the other one may be taken as the frequency, ω. Since this does not appear in the potential function, we find

$$\frac{\mathrm{d}\omega}{\mathrm{d}t} = -\frac{\partial \phi}{\partial \omega} = 0,$$

* I am indebted to Dr Manuel Berondo, Institute of Physics, National Autonomous University of Mexico, for introducing me to this work.

whence $\omega = \text{constant}$, determined by the factors described above. The function equation (7.12) belongs to the same family as that used to describe gas condensation, equation (7.11), viz. the cusp or Riemann–Hugoniot catastrophe, with the difference that symmetry considerations concerning the oscillation result in one of the unfolding parameters, u_2, being taken equal to zero. As a result, there is no shock wave in the potential function but only a bifurcation controlled by the parameter u_1, interpreted as a phase transition.

An interesting extension of the dynamical description of the system is possible which allows one to include random fluctuations in the process by the simple expedient of adding a term $c\dot{r}^2$ (where $\dot{r} = dr/dt$) to the potential function equation (7.12), where c is a parameter determining the size of the fluctuations. It is also possible to derive a probability distribution function for the system, which is obtained as the solution of a Fokker–Planck equation, providing certain conditions on the process are satisfied. This procedure is described by Graham (1973), and I want only to mention that the equilibrium distribution function, which is the analogue of that used in statistical mechanics, has the familiar form

$$P = Ne^{-\phi} \tag{7.13}$$

where N is a normalizing constant. With this function one may obtain any required statistical properties of the synchronization process. The similarity between this procedure and that of statistical mechanics is now obvious, despite the fact that we are dealing with a system which is far from thermodynamic equilibrium. The extended function $\phi(r, \dot{r})$ evidently plays the role of a Hamiltonian in this treatment. Observe that in this situation there is no macroscopic variable analogous to the temperature which enters equation (7.8) in the form $\beta = 1/kT$, arising from the assumption that the transition from non-excitability in the membrane occurs close to a condition of thermodynamic equilibrium. Such a constraint is dispensed with in deriving equation (7.13) for the system which is far from thermodynamic equilibrium, although other constraints relating to the Markoff nature of the fluctuations and stationarity of parameter values must be satisfied. What I think is of considerable importance is the possibility of using the potential functions of catastrophe theory within this generalized dynamic context, and thus developing a macroscopic description of biological process which shares many features with equilibrium thermodynamics, but is rigorously applicable to open system behaviour. It is the absence of descriptive potential functions from the Glansdorff–Prigogine treatment of non-equilibrium processes which I believe to be its chief weakness, leaving one with essentially no more than the procedures of stability theory to handle the complexities of biological processes.

The descriptive procedure using the potential function equation (7.12) or its stochastic extension may be applied to other population rhythms, such as the diurnal rhythm of mitotic activity in mammalian epidermis with adrenaline concentration as the synchronizing periodicity (Bullough and Laurence, 1964), or the occurrence of a circadian rhythm in a population of organisms as a result of entrainment to a periodic forcing function or Zeitgeber. In each instance there is an external parameter controlling the appearance of the collective rhythm, whose frequency is determined by a variety of factors including the periodicity of the forcing function. It is of interest to consider in what way it is necessary to modify this description when considering the cycle of cell growth and division in single cells. In this case growth is again under the control of an external parameter, the concentration of a rate-limiting nutrient in the culture medium, for example, but the frequency of the cell cycle is not a result of any periodic forcing function and is a part of the system dynamics itself. Let us now see how one may describe this in phenomenological terms.

Let us take the view that the transition from the non-growing to the growing state in a single cell involves the appearance of a collective or co-operative periodicity among the different functional control circuits or cognitive units of the growing cell, this transition occurring when the rate-limiting nutrient concentration rises above a particular threshold value. No details will be considered regarding the microscopic or molecular dynamics of the interacting processes underlying the appearance of this macroscopic periodicity, observed as the growth and division of the cell and its progeny. The potential function will, as before, be used to describe the dynamical phase transition, and will contain two variables, a mode amplitude, r, and a frequency, ω. The amplitude measures the coherence of the collective oscillation, while ω gives its frequency, both variables being required in this case. This leads us to consider Thom's potential functions with two variables, called the umbilics. The first of these is the hyperbolic umbilic, with potential function

$$\phi(r, \omega) = r^3 + \omega^3 - \lambda r\omega - ur - v\omega. \tag{7.14}$$

This gives for the dynamics the equations

$$\begin{aligned} \frac{\mathrm{d}r}{\mathrm{d}t} &= -\frac{\partial \phi}{\partial r} = -3r^2 + \lambda\omega + u \\ \frac{\mathrm{d}\omega}{\mathrm{d}t} &= -\frac{\partial \phi}{\partial \omega} = -3\omega^2 + \lambda r + v \end{aligned} \tag{7.15}$$

These equations may be taken to define a limit cycle of amplitude r and frequency ω, both of which are stable to perturbation. It is convenient to take $u = 0$ and to choose λ as the controlling parameter, which is a function of the limiting nutrient concentration. The considerations of Chapter 4 regarding the growth of cells in controlled conditions suggest that we take

$$\lambda(s) = \frac{\mu_m s}{K_m + s} - k \qquad (7.16)$$

where the first part is the familiar Monod expression for growth of cells on limiting substrate, s, while the constant k determines the critical value of s, call it s_c, at which growth begins. This value is the root of the equation $\lambda(s) = 0$, which gives

$$s_c = \frac{kK_m}{\mu_m - k}.$$

The steady state values of r and ω are defined by

$$\left.\begin{aligned} -3\bar{r}^2 + \lambda\bar{\omega} &= 0 \\ -3\bar{\omega}^2 + \lambda\bar{r} + v &= 0 \end{aligned}\right\}. \qquad (7.17)$$

Evidently if $\lambda = 0$, $\bar{r} = 0$, implying that there is then no collective or coherent oscillation in the system so that there is no growth. For $\lambda < 0$ there is no real solution in $\bar{r}$, so the same interpretation holds. Oscillations in individual control circuits within a non-growing cell may still go on, but the pattern of interactions will be such that no collective oscillatory motion occurs throughout the cell. An interesting possibility suggested by this view of the transition from no growth to growth is that there may be certain frequencies of epigenetic or metabolic variables which are incompatible with collective periodic motion of the whole cell, so that by driving intracellular variables at certain frequencies (say by a periodicity of pH in the culture) it may be possible to prevent growth as a result of incoherence in the dynamics.

The growth rate of a cell will clearly be directly proportional to the frequency of the collective oscillation in the control circuits, so we will take

$$\mu = C\omega \qquad (7.18)$$

where μ is the growth rate and C is a constant. By the same procedures as those mentioned above concerning the stochastic extension of these phenomenologically-defined processes, one may obtain a probability distribution which allows one to determine the statistical properties of the system by adding to the potential function the terms $c\dot{r}^2$ and $d\dot{\omega}^2$, c and d being constants. Then one can calculate means and variances and higher moments, if required. The mean values of r and ω will be the same as their steady-state values, so we find from equation (7.18) that

$$\bar{\mu} = C\bar{\omega}.$$

Now if we take $v = 0$, we find that $\bar{\omega} = \lambda/3$ so that

$$\bar{\mu} = \frac{C\lambda}{3} - C'\left(\frac{\mu_m s}{K_m + s} - k\right). \tag{7.19}$$

An empirically established expression which is used to describe the growth of cells in chemostat culture is

$$\mu = Y(Q - M) \tag{7.20}$$

where Y is the yield coefficient defined in Chapter 4, Q is the rate of uptake of the limiting nutrient and M is the maintenance constant, the amount of nutrient required per unit time to maintain the cell in a viable state without growth. This quantity was omitted from the considerations of Chapter 4 because it is a small correction term, but it is required for an accurate description of the growth process. Evidently (7.19) can be easily identified with (7.20), so that the model used recovers an established expression. This results from the definition of λ as an unfolding parameter of the potential function and the manner of its dependence upon limiting nutrient concentration.

An important difference between the description of cell synchronization on the one hand and the growth process on the other, both viewed as phase transitions describable by phenomenological potential functions, is that in the latter case both amplitude and frequency are treated as variables with stable dynamics, as described by equations (7.15). These equations have the right properties for the description of a system with limit cycle type stability, since a perturbation which increases the amplitude from the steady state value results in a decrease in the rate of change of the amplitude but an increase in the rate of change of the frequency; and conversely. It is this type of stability that is inferred from experiments such as those of Kauffman and Wille (1975) on the mitotic cycle in *Physarum*. The variables r and ω used above are mean values of the radial co-ordinate and the angular velocity, $d\theta/dt$, over a complete cycle of the oscillation.

If one wishes to extend the analysis to a situation in which there are interactions between growing cells such that spontaneous synchronization develops, then it is necessary to introduce interaction terms between cells. Suppose that the dynamics of each cell is described by a potential function $\phi_i(r_i, \omega_i)$ like equation (7.14), in which the subscript refers to the ith cell in the population. In addition there are interactions which may be expressed as follows, giving a potential function for the population in the form

$$\phi(r_1, r_2, \ldots, r_n; \omega_1, \omega_2, \ldots, \omega_n) = \sum_{i=1}^{n} (r_i^3 + \omega_i^3 - \lambda_i r_i \omega_i - u_i v_i - v_i \omega_i) - \tfrac{1}{2} \sum a_{ij} \cos(\omega_i - \omega_j) \qquad (7.21)$$

$$[a_{ij} = a_{ji}].$$

This gives the dynamical equations

$$\frac{dr_i}{dt} = -3r_i^2 + \lambda_i \omega_i + u_i$$

$$\frac{d\omega_i}{dt} = -3\omega_i^2 + \lambda_i r_i + v_i - \sum_{j=1}^{n} a_{ij} \sin(\omega_i - \omega_j).$$

The interaction terms tend to draw the cells into a condition of synchrony in which $\omega_i = \omega_j$. By extending the dynamics in the usual manner to include fluctuations, it is possible to describe a condition of noisy synchrony in the population.

These examples illustrate the use of phenomenological potential functions for the description of non-linear, complex processes in biological systems which manifest themselves collectively or co-operatively. No sharp discontinuities of state such as gas condensation or the transition from non-excitability to excitability described earlier have been included in the cases of temporal organization considered above, so that the domains of occurrence of 'shock waves' in the potential functions have been excluded. These are, of course, available for the description of sudden changes of state in terms of jumps from one steady-state to another in accordance with a convention such as Maxwell's. A possible example of this type of behaviour may be seen in the development of *Dictyostelium discoideum* where the frequency of a pacemaker changes rather suddenly from about 1 pulse/10 min to about 1 pulse/5 min, as reported by Durston (1974). Such a discontinuity of frequency may be described in terms of a fold region in the potential function equation (7.14), leading to a jump from one part of the surface to another along the frequency axis, as occurs in the case of gas condensation along the axis of volume. Temporal discontinuities of this type are not yet commonly known as aspects of the developmental process, but they are certainly a significant aspect of behaviour, such as the sudden change of frequency associated with changes of modality such as the transition from walking to running. Some observations on the relationships between the processes underlying embryogenesis and behaviour will be considered below.

Discontinuities of state along a spatial axis arise in the same way as those in time, but in this instance one of the unfolding parameters of the potential function represents space. Zeeman (1974) has given a very complete description of this procedure in his application of the cusp catastrophe to the cases of amphibian gastrulation and to culmination in the slime mould. Along the spatial axis cell state varies in a monotonic manner, determined by a gradient of a morphogen, so that the analogue of the critical temperature in the gas condensation case is a critical value of the morphogen, occurring at some point of space beyond which discontinuous change of cell state can occur. A boundary between two regions with cells in different states then arises, so that spatially-ordered differentiation domains are generated, such as those separating presumptive mesoderm from ectoderm, or cartilage from muscle. A particularly interesting feature of Zeeman's description is his deduction of how waves of state change travel through the developing tissue as the discontinuity unfolds, stopping at boundaries which are determined by the morphogen gradient. The functions used to describe these discontinuities can give some information about the geometry of the discontinuities, but none about the overall shape of the tissue undergoing the controlled discontinuity, so the procedure is obviously limited. What is needed for this is a genuinely global theory of the generation of form in living systems which requires, I believe, a quite different approach to biological order from the current preoccupation with local interactions and intermolecular forces of the type we have been concerned with throughout this book. However, the possibility of giving rigorous mathematical descriptions of complex state transitions and embedding these within the framework of far from equilibrium dynamic processes with macroscopic probability distribution functions, represents a very significant advance in our ability to handle some basic qualitative features of epigenetic and morphogenetic processes. Furthermore, it is possible to extend these procedures to include propagating activity waves as travelling discontinuities, so that the general picture of embryogenesis presented in Chapters 5 and 6 becomes amenable to mathematical description and analysis. This can then be embedded in the general framework of inverse limit systems as described at the beginning of the chapter, to provide a cognitive context for the whole developmental process. In concluding this chapter, I shall consider a few of the more general implications of this context.

Generative Processes in Cognitive Systems

The appearance of collective dynamic modes and spatial discontinuities within a complex system, arising from patterns of interaction between units at some level in the system and describable as phase transitions, results in a greatly simplified descriptive language whereby the system informs itself of its own state and determines subsequent transformations. These modes are measures of cognitive performance in so far as they signal the emergence of useful order in the system, expressing correlated behaviour and acting as inputs and/or constraints for further development. Embryogenesis may then be seen as the progressive, orderly manifestation of the knowledge which is latent in the egg. This knowledge is of environment and self, resulting in an organism which can survive and reproduce. The inverse limit system of Figure 7.1 thus comes into being as one level after another unfolds and as interactions within and between levels generate spatial and temporal organization by the controlled emergence of phase transitions which produce ordered cell states in space and collective rhythmic behaviour in time. It is, of course, necessary to think of these processes occurring in an interdependent and correlated manner. For example, in the wave model a phase transition to an excitable state on the membrane (space) generates rhythmic propagating activity waves (time and space) producing a gradient (space) which determines rates of cell differentiation (time) leading to ordered spatial discontinuities. It is clear that in such a description it is artificial and unnecessary to separate out the concepts of positional information and its interpretation, since there is a continuity of the four-dimensional unfolding process. One should seek models which capture this four-dimensional quality of embryogenesis and at the same time recognize the fact that the developing organism is undergoing transformations which are constrained by the phylogenetic history of the species and the necessity to generate a dynamic structure capable of surviving and reproducing in its environment. These constraints determine the knowledge or the useful descriptions which the embryo contains in its own structure in the form of homeostatic control circuits, membranes which can become excitable, kinetic interactions which lead to coherent rhythmic activity and propagating waves, switching-type gene control circuits, etc. The cognitive language of the developing organism is the set of commands which order these processes in space-time. These commands are usually concentrations of substances which elicit either continuous responses, as in a control circuit, or discontinuous behaviour. The latter arises when concentrations exceed the threshold points characterized by the critical values of particular controlling or unfolding parameters, initiating specific phase transitions as described above.

As well as interactions between units at one level of the unfolding hierarchy, there will be interactions occurring in both directions between different levels, averaging and correlating in one direction, controlling and constraining in the other. This picture of a self-organizing system with an overall continuous, stable dynamic within which controlled, spatial and temporal ordering discontinuities occur begins to satisfy the requirements of the analogue-digital, or synthetic-analytic duality which I proposed (Goodwin, 1968, 1970) as a basic requirement for any adequate model of biological processes. Pattee (1974) has more recently suggested that the continuous-discrete or dynamic-symbolic property of biological systems may be the root of intelligent behaviour, a position very similar to the cognitive view of organisms being advanced here.

In addition to interactions within the developing organism, described above, one must take account also of its interactions with the outside world, which become richer as the embryo approaches maturity. With the development of the sense organs in the animal, specific variables enter the system and act as controllers of behaviour, initiating switches and transitions of modality in a manner qualitatively similar to those operating in the developmental process. It is at this point that an important feature of the cognitive description of organisms emerges. Implicit in this view is the notion of the continuity of the organizing principles in development and behaviour. The latter is conventionally regarded as part of the cognitive domain, since the activities of the adult organism are recognized to be ordered by knowledge, whether learned or instinctive. Behaviour may be and has been described as continuous with higher cognitive activity such as problem-solving, speech, artistic activity, etc., so that a cognitive view of organismic activities in fact suggests a continuity of structural and functional principles throughout all levels of the organic domain, uniting animate nature and mind in a continuum. This is a deliberate consequence of the view advanced in this chapter and one which I shall explore briefly now. I have indicated elsewhere (Goodwin, 1973, 1975) some of its implications as regards behaviour.

Let us start by recalling some observations made in Chapter 5. In the development of *Dictyostelium*, periodic waves of acrasin release propagating from pacemaker centres constitute the basic signalling mechanism for cell aggregation, the first stage of development in the organism. After aggregation has been completed and the slug has formed, the pacemaker cells continue their activity, now functioning as the tip of the migrating slug and controlling its movement (Cohen and Robertson, 1972). During the final phase of the developmental process, the erection of the fruiting body, periodic upward movement of the sorocarp is again observed, presumably under the control of the same cells acting as an organizing centre. In each of these three processes, the same basic activity results in three different phases of the developmental process, the differences resulting from the fact that the context within which the activity is expressed changes. First there are dispersed amoebae; then a continuous mass of amoebae with cell–cell contact and active production of a polysaccharide sheath around the mass; and finally cells ordered spatially into two types, stalk and spore, the former undergoing transformation to a vacuolated type with a rigid wall. The mechanisms operating during these phases are far from being understood in detail, but the point I wish to extract is the context-sensitivity of developmental processes: the same events occurring in different situations have quite distinct morphological consequences.

There are many examples of such behaviour in developing systems, but the case of *Tubularia* described in Chapter 5 is particularly striking. After cutting off the hydranth, a series of quiescent and pulsatile phases were observed in the presumptive hydranth region, the latter showing periodic propagating contraction waves with periods of the order of 8–10 minutes. The final phase of regeneration in this organism was seen to be the pulsatile emergence of the fully-differentiated hydranth from the perisarc, followed by the vigorous movements of the tentacles which are the means whereby food is brought into the hypostome. In this case the periodic activity of a developmental organizer appears to continue in the adult, at higher frequency, and presumably from specialized cells, as a behavioural organizer for feeding movements. This supposition is strengthened by the studies of Passano and McCullough (1965) on the system of pacemakers which control behavioural movements in *Hydra*. These are located at various positions in the animal such as the hypostome, base of the tentacles, peduncle, etc. Developmental pacemakers have the role of generating embryological fields and organizing morphogenetic transformations. Behavioural pacemakers organize and co-ordinate the movements of the adult body. The developmental field is replaced by a behavioural field whose function is not the morphological transformation of the organism but the transformation of the relationship between the organism and its environment: movement to a new location, for example, or transfer of food from the outside to the inside of the organism. This change of activity results from a change of context within which the pacemakers operate and also changes in their own structure. Thus the nerve net of hydroids acts as a communication network, the relatively specialized cells of which it is composed conducting the activity waves without setting up spatial gradients of morphogens. This requires some simple molecular uncoupling during differentiation, and some specialization of the membrane for rapid conduction of activity waves as well as higher frequency of pacemaker activity (1–10 per minute in *Hydra*). However, the simple developmental process of coupling activity waves and gradient formation is available for any structural changes required in association with behavioural responses, such as changes in network connectivity accompanying learning. This does not seem to be an aspect of hydroid behaviour, but this mechanism may be the one operating in association with learning and memory in higher organisms.

Examples illustrating recursive features and the continuity of development and behaviour such as the cases of *Dictyostelium* and *Tubularia* described earlier can be extended up the phylogenetic scale. However, I think the point has now been made that it is reasonable to see a continuity of certain organizing principles from evolutionary and developmental processes to adult behaviour. Of particular interest is the concept of context sensitive activity mentioned above in relation to aspects of embryogeneses. As used in the study of language, context-sensitivity has the same meaning. Consider, for example, the sentence 'I saw the dog'. The context tells us the meaning of saw and we do not imagine someone sawing through a dog. An interesting feature of the linguistic case is the use of the concept, 'meaning'. We would not normally say that the meaning of an aggregation signal is the summoning of amoebae to form a slug, although this would be a perfectly consistent way of talking about its significance. It is simply not fashionable to talk in semantic terms in reference to slime mould aggregation, although we use this form all the time when we talk about cognitive processes at the psychological level. Meaning here is signification, and it is the context of a word, the pattern of its interactions with other words in a sentence, which determines this, just as it is the pattern of interactions between cAMP and slime mould cells which determines what activity results, hence its significance.

Now we know that in language it is possible to have grammatically (syntatically) correct sentences which have no meaning within the domain of experience, such as 'The apple ate the boy'. This has no experiential meaning because it does not correspond to a realizable situation in the world as we know it. The same thing can occur in developmental processes. For example, the phenomenon of exogastrulation in an amphibian embryo is perfectly consistent with all the local developmental constraints (syntactical rules) governing gastrulation, such as the movement of ectodermal cells at the lip of the blastophore, the polar structure of the bottle cells which initiate gastrulation, tangential contractions of cells in the grey crescent region, etc. But exogastrulation is developmentally meaningless: it cannot be fitted into the set of events which constitute normal development: it is a cul-de-sac. This interpretation emphasizes the fact that meaning is relative to a particular domain of what is called normal experience, referring only to possible processes. An additional freedom which language has over developmental processes is that linguistic statements can refer to virtual as well as to actual events, hence to imaginable possibilities. The embryo cannot do this, since it must actually realize any cognitive process it is engaged in, having no symbolic domain of operation divorced from action. All 'statements' in developmental language are commands or algorithms. This is precisely the point made by Waddington (1972), who said: 'To a biologist, therefore, a language is a set of symbols, organized by some sort of generative grammar, which makes possible the conveyance of (more or less) precise commands for action to produce effects on the surroundings of the emitting and recipient entities . . . . It is language in this sense—not as a mere vehicle of vacuous information—that I suggest may become a paradigm for the theory of General Biology'. These commands operate within a system constantly subjected to the necessity to survive, to preserve its order and organization: the evolutionary process requires organisms to be practical, to connect with 'reality'. In terms of the definition of a cognitive system introduced at the beginning of this chapter, language functions cognitively and is a vehicle for knowledge only in so far as it conveys useful descriptions of aspects of the organism or its environment. Then it acts as part of the evolutionary process.

One fruitful result that could come from this juxtaposition of embryogenesis and language is a clarification of the biological roots of generative processes. In linguistics, the aim of a generative grammar is relatively modest: to devise a set of operations whereby the surface structure of language is correctly generated. In epigenesis, a generative theory must give some prescription whereby the adult form is generated step by step from the egg. The meaning of the knowledge contained within the egg becomes manifest gradually throughout development until the final details of its surface structure are fully unfolded and the organism enters into a relationship with its environment. The orderly, rule-obeying embryological process is the analogue of a generative grammar, but in this case the process originates at its root; the germ of the organism, the egg with its store of potential knowledge. What is the germ of a sentence, of an action, of a myth? What is its order of unfolding? Of course language and symbol systems generally, have particular constraints which determine many aspects of their surface structure, and these differ markedly from those operating in embryos. But the possibility glimpsed here is to see these as examples of cognitive processes generating their own particular morphologies, whether embryological, behavioural, linguistic, mythological, or other. These may all share similar underlying dynamical features, may all be isomorphic with one another at some deep level so that comparative studies among these domains become reciprocally illuminating. One may then indeed have a foundation for a theoretical biology uniting these areas of investigation, all of which are embedded in the evolutionary process.

A final word concerns the nature of the evolutionary process relative to the cognitive perspective being presented here. It is reasonable to suggest from this perspective that a process capable of generating useful descriptions of aspects of the world and storing them in accessible form manifests intelligence, the capacity to learn. For example, we can say that when someone solves a problem he is generating a useful description of that part of the world defined by the problem, and his ability to generate this description and to store it in accessible form (to remember it) is a measure of his intelligence. By this definition the evolutionary process manifests intelligence. It seems to me undeniable that the products of this process, organisms, are indeed the results of the working of an intelligent system and the above definition of intelligence provides a formal, explicit and self-consistent way of stating this intuition. Organisms themselves need not, of course, be intelligent: a cognitive system operates on the basis of useful descriptions of aspects of the world, but need not have the capacity to generate and store them. It is clear that a cellular slime mould, for example, is not intelligent, that it cannot learn. This ability emerges at some level of biological organization and organisms then share with the evolutionary process the property of intelligence, although the detailed mechanisms they employ in manifesting this property are quite different. Whereas the trial and error procedure in the evolutionary learning process involves mutation and recombination of the hereditary elements together with natural selection, a process requiring many generations, in an organism the generation of useful descriptions involves the recombination of the elements of behavioural strategies and their testing against the environmental situation, a process which takes place within the organism's own life span.

These brief and perhaps provocative suggestions about possible relationships between evolutionary and organismic processes require detailed study and the development of a rigorous model to demonstrate both their plausibility and their utility. Loose analogies can be very misleading, but genuine homologies and isomorphisms can be of the greatest value in allowing reciprocal enrichment across fields of study which have become isolated one from the other. Of even greater value is the possibility of grasping the overall evolutionary process, in which all aspects of nature and culture are involved, as a cognitive unfolding which is based upon the operation intelligence. Then we might begin to find our way through the labyrinthine garden of forking paths which the mind creates as it generates the forms experienced in space and in time.

REFERENCES

Adler, H. I., Fisher, W. D., Cohen, A. and Hardigree, A. A. (1967). Miniature *Escherichia coli* cells deficient in DNA. *Proc. natn. Acad. Sci. U.S.A.* **57,** 321.

Aldridge, J. F. and Pavlidis, T. (1976). Clocklike behaviour of biological clocks. *Nature, Lond.* **259,** 343.

Apter, M. J. (1966). "Cybernetics and Development", Pergamon Press, Oxford.

Aschoff, J. (ed.) (1965) "Circadian Clocks," North-Holland, Amsterdam.

Atkinson, D. E., Hathaway, J. A. and Smith, E. C. (1965). Kinetic order of the yeast diphosphopyridine nucleotide isocitrate dehydrogenase reaction and a model for the reaction. *J. biol. Chem.* **240,** 2682.

Azurnia, R. and Lowenstein, W. R. (1971). Intercellular communication and tissue growth. V. A cancer cell strain that fails to make permeable membrane junctions with normal cells. *J. Membr. Biol.* **6,** 368.

Barbera, A. J., Marchase, R. B. and Roth, S. (1973). Adhesive recognition and retinotectal specificity, *Proc. natn. Acad. Sci. U.S.A.* **70,** 2482.

Bard, J. and Lauder, I. (1974). How well does Turing's theory of morphogenesis work? *J. theor. Biol.* **45,** 501.

Barenska, J. and Wlodaver, P. (1969). Influence of temperature on the composition of fatty acids and on lipogenesis in frog tissues. *Comp. Biochem. Physiol.* **28,** 553.

Bateman, A. E. (1974). Cell specificity of chalone-type inhibitors of DNA synthesis released by blood leucocytes and erythrocytes. *Cell Tiss. Kinet.* **7,** 451.

Bateman, A. E. and Goodwin, B. C. (1976). Erythrocyte chalone; demonstration of an inhibitor of DNA synthesis in erythrocyte-conditioned medium, action on foetal erythroblasts, and its separation from haemoglobin. *Biomedicine* (in the press).

Becker, R. O. (1972). Stimulation of partial limb regeneration in rats. *Nature, Lond.* **235,** 109.

Beckworth, J. R. (1967). Regulation of the lac operon. *Science, N.Y.* **156,** 597.

Beloussov, L. V., Badenko, L. A., Katchurin, A. L. and Kunilo, L. F. (Filacheva) (1972). Cell movements in morphogenesis of hydroid polyps. *J. Embryol. exp. Morph.* **27,** 317.

Berridge, M. J. (1975). The interaction of cyclic nucleotides and calcium in the control of cellular activity. *Adv. Cyclic Nucleotide Res.* **6,** 1.

Bertalanffy von, L. (1952). "Problems of Life," C. A. Watts, London.

Beug, H., Katz, F. E., Stern, A. and Gerisch, G., (1973). Membrane sites in aggregating *D. discoideum* cells by use of tritiated univalent antibody. *Proc. natn. Acad. Sci. U.S.A.* **70,** 3150.

Bohn, H. (1974). Pattern reconstitution in abdominal segment of *Leucophaca maderae* (Blattaria). *Nature, Lond.* **248,** 608.

Bonner, J. T., Barkley, D. S., Hall, E. M., Konijn, T. M., Mason, J. W., O'Keefe, G. and White, B. P. (1969). Acrasin, acrasinase, and the sensitivity to acrasin in *Dictyostelium discoideum*. *Devl Biol.* **20,** 72.

Borek, C., Higashira, S. and Lowenstein, W. R. (1969). Intercellular communication and tissue growth. IV. Conductance of membrane junctions of normal and cancerous cells in culture. *J. Membr. Biol.* **1,** 274.

Brachet, J. (1964). The role of nucleic acids and sulfhydryl groups in morphogenesis (amphibian egg development, regeneration in *Acetabularia*). *In* "Advances in Morphogenesis" (eds M. Abercrombie and J. Brachet), Vol. 3, p. 247, Academic Press, New York and London.

Brachet, J. and Bonotto, S. (1970). "Biology of *Acetabularia*". Academic Press, New York and London.
Bradbury, E. M., Inglis, R. J. and Matthews, H. R. (1974a). Control of cell division by very lysine-rich Histone (FI) phosphorylation. *Nature, Lond.* **247,** 257.
Bradbury, E. M., Inglis, R. J., Matthews, H. R. and Langam, T. A. (1974b). Molecular basis of control of mitotic cell division in eukaryotes. *Nature, Lond.* **249,** 553.
Bremel, R. D. and Weber, A. (1972). Co-operation within actin filament in vertebrate skeletal muscle. *Nature, New Biology* **238,** 97.
Bretscher, M. S. (1971). Principal glycoprotein on the surface extends into the interior in human erythrocytes, *Nature, New Biology* **231,** 229.
Brewer, E. N. and Rusch, H. P. (1968). Effect of elevated temperature shocks on mitosis and on the initiation of DNA replication in *Physarum polycephalum. Expl Cell. Res.* **49,** 79.
Britten, R. J. and Davidson, E. H. (1969). Gene regulation for higher cells: a theory. *Science, N.Y.* **165,** 349.
Brout, R. (1965). "Phase Transitions," Benjamin, New York.
Brown, J. M. and Berry, R. F. (1968). The relationship between diurnal variation of the number of cells in mitosis and of the number of cells synthesizing DNA in the epithelium of the hamster cheek pouch. *Cell Tiss. Kinet.* **1,** 23.
Bruce, V. (1965). Cell division rhythms and the circadian clock. *In* "Circadian Clocks" (ed. J. Aschoff), p. 125. North-Holland, Amsterdam.
Bryant, P. J. (1971). Regeneration and duplication following operations *in situ* on the imaginal disc of *Drosophila melanogaster. Devl Biol.* **26,** 637.
Bryant, P. J. (1975a). Pattern formation in the imaginal wing disc of *Drosophila melanogaster* fate map, regeneration, and duplication. *J. exp. Zool.* **193,** 49.
Bryant, P. J. (1975b). Regeneration and duplication in imaginal discs. *In* "Cell Patterning" (eds R. Porter and J. Rivers), Ciba Foundation Symposium, Vol. 29, p. 71, Elsevier, Amsterdam.
Bullough, W. S. (1962). The control of mitotic activity in adult mammalian tissues. *Biol. Rev.* **37,** 307.
Bullough, W. S. (1975a). Mitotic control in adult mammalian tissues. *Biol. Rev.* **50,** 99.
Bullough, W. S. (1975b). Chalones and cancer. *In* "Host Defense against Cancer and its Potentiation", (eds D. Mizano *et al.*), p. 317, University of Tokyo Press, Tokyo/University Park Press, Baltimore, U.S.A.
Bullough, W. S. and Laurence, E. B. (1960). The control of epidermal mitotic activity in the mouse. *Proc. R. Soc.* B. **151,** 517.
Bullough, W. S. and Laurence, E. B. (1961). Stress and adrenaline in relation to the diurnal cycle of epidermal mitotic activity in adult male mice. *Proc. R. Soc.* B. **154,** 540.
Bullough, W. S. and Laurence, E. B. (1964). Mitotic control by an internal secretion: the role of the chalone–adrenaline complex. *Expl Cell Res.* **33,** 176.
Bullough, W. S. and Laurence, E. B. (1968). Epidermal chalone and mitotic control in the VX2 epidermal tumour. *Nature, Lond.* **220,** 134.
Bünning, E. (1967). "The Physiological Clock", Springer-Verlag, New York and Heidelberg.
Bünning, E. and Moser, I. (1972). Influence of valinomycin on circadian leaf movements of *Phaseolus. Proc. natn. Acad. Sci. U.S.A.* **69,** 2732.
Burstein, C., Cohn, M., Kepes, A. and Monod, J. (1965). Rôle du lactose et de ses produits métaboliques dans l'induction de l'opéron lactose chez *Escherichia coli. Biochim. biophys. Acta* **95,** 634.
Burton, A. C. (1971). Cellular communication, contact inhibition, cell clocks, and cancer: the impact of the work and ideas of W. R. Lowenstein. *Perspect. Biol. Med.* **14,** 301.
Burton, A. C. and Canham, P. B. (1973). The behaviour of coupled biochemical oscillations as a model of contact inhibition of cellular division. *J. theor. Biol.* **34,** 555.
Campbell, R. D. (1967). Tissue dynamics of steady state growth in *Hydra littoralis*:
I. Patterns of cell division. *Devl Biol.* **15,** 487;
II. Patterns of tissue movement. *J. Morph.* **121,** 19.
Cashel, M. and Gallant, J. (1969). Two compounds implicated in the function of the RC gene of *Escherichia coli. Nature, Lond.* **221,** 838.

Caveney, S. (1973). Stability of polarity in the epidermis of a beetle, *Tenebrio molitor L. Devl Biol.* **30,** 321.

Chandrashekaran, M. K. (1967). Studies on phase-shifts in endogenous rhythms. *Z. vergl. Physiol.* **56,** 154.

Changeux, J.-P. (1963). Allosteric interactions on biosynthetic L-threonine deaminase from *E. coli* K 12. *Cold Spring Harb. Symp. quant. Biol.* **28,** 497.

Child, C. M. (1941). "Patterns and Problems of Development", Chicago University Press.

Chomsky, N. (1968). "Language and Mind", Harcourt Brace, New York.

Chomsky, N. (1972). "Problems of Knowledge and Freedom", p. 26, Fontana/Collins.

Chung, Shin-Ho and Cooke, J. (1975). Polarity of structure and of ordered nerve connections in the developing amphibian brain. *Nature, Lond.*, **258,** 126.

Cohen, M. H. (1972). Models of clocks and maps in developing organisms. *In* "Lectures on Mathematics in the Life Sciences" (ed. J. D. Cowan) Vol. III, p. 3, American Mathematical Society, Providence, Rhode Island.

Cohen, M. H. and Robertson, A. (1971a). Wave propagation in the early stages of aggregation in cellular slime molds. *J. theor. Biol.* **31,** 101.

Cohen, M. H. and Robertson, A. (1971b). Chemotaxis and the early stages of aggregation in cellular slime molds. *J. theor. Biol.* **31,** 119.

Cohen, M. H. and Robertson, A. (1972). Differentiation for aggregation in the cellular slime molds. *In* "Proceedings of the First International Conference on Cell Differentiation" (eds R. Harris and D. Viza), p. 35, Munkesgaard, Copenhagen.

Cohlberg, J. A., Piget, V. P. and Schackman, H. K. (1972). Structure and arrangement of the regulatory subunits in aspartate transcarbamylase. *Biochemistry, N.Y.* **11,** 3396.

Cohn, M. and Torriani, A. M. (1952). Immunochemical studies with the β-galactosidase and structurally related proteins of *Escherichia coli. J. Immun.* **69,** 471.

Cold Spring Harb. Symp. quant. Biol. (1960) **25.** Biological Clocks.

Cone, C. D. (1969). Some theoretical aspects of intercellular bridges as a potential mechanism of cancerous proliferation. *J. theor. Biol.* **22,** 365.

Cooke, J. (1972). Properties of the primary organization field in the embryo of *Xenopus laevis*:
I. Autonomy of cell behaviour at the site of initial organizer function. *J. Embryol. exp. Morph.* **28,** 13;
II. Positional information for axial organization in embryos with two head organizers. *J. Embryol. exp. Morph.* **28,** 27;
III. Retention of polarity in cell groups excised from the region of the early organizer. *J. Embryol. exp. Morph.* **28,** 47.

Cooke, J. (1973). Morphogenesis and regulation in spite of continued mitotic inhibition in *Xenopus* embryos. *Nature, Lond.* **242,** 5392.

Cooke, J. (1975a). Control of somite number during morphogenesis of a vertebrate, *Xenopus laevis. Nature, Lond.* **254,** 196.

Cooke, J. (1975b). Experimental analysis and a model of the control of somite formation in *Xenopus laevis. In* "Developmental Biology" (U.C.L.A., Squaw Valley Winter Conference), Walter Benjamin, Menlo Park, California.

Crombrugghe de, B., Chero, B., Gotterman, M., Paston, I., Varmus, H. E., Emmer, M. and Perlman, R. L. (1971). Regulation of lac mRNA synthesis in a soluble cell-free system. *Nature, New Biology* **230,** 37.

Crick, F. H. C. (1971). General model for the chromosomes of higher organisms. *Nature, Lond.* **234,** 25.

Cummings, F. W. (1975). A biochemical model of the circadian clock. *J. theor. Biol.* **55,** 455.

Cummings, F. W. (1976). On the metabolic origins of the circadian period. (to be submitted for publication).

David, C. N. and Campbell, R. D. (1972). Cell cycle kineties and development of *Hydra attenuata. J. Cell Sci.* **11,** 557.

Davson, H. and Danielli, J. F. (1952). "The Permeability of Natural Membranes", Cambridge University Press.

Degn, H. and Meyer, D. (1969). Theory of oscillations in peroxidase catalyzed oxidation reactions in open systems. *Biochim. biophys. Acta* **180,** 291.

Denbigh, K. G., Hicks, M. and Page, F. M. (1948). The kinetics of open reaction systems. *Trans. Faraday Soc.* **44,** 479.
Donachie, W. D. and Begg, K. J. (1970). Growth of the bacterial cell. *Nature, Lond.* **227,** 1220.
Donachie, W. D., Jones, N. C. and Teather, R. (1973). The bacterial cell cycle. *Symp. Soc. gen. Microbiol.* **23,** 9.
Driesch, H. (1908). "Science and Philosophy of the Organism", A. and C. Black, London.
Duclaux, E. (1899). "Traité de Microbiologie", Masson et Cie, Paris.
Durston, A. J. (1973). *Dictyostelium discoideum* aggregation fields as excitable media. *J. theor. Biol.* **42,** 483.
Durston, A. J. (1974). Pacemaker activity during aggregation in *Dictyostelium discoideum. Devl Biol.* **37,** 225.
Duysens, L. N. M. and Amesz, J. (1957). Fluorescence spectrophotometry of reduced phosphopyridine nucleotide in intact cells in the near-ultra-violet and visible region. *Biochim. biophys. Acta* **24,** 19.
Edmunds, L. N. (1971). Persistent circadian rhythm of cell division in *Euglena*: some theoretical considerations and the problem of intercellular communication. *In* "Biochronometry", p. 594. National Academy of Sciences, Washington.
Edmunds, L. N. and Cirillo, V. P. (1974). On the interplay among cell cycle, biological clock, and membrane transport control systems. *Int. J. Chronobiol.* **2,** 233.
Ehret, C. F. and Trucco, E. (1967). Molecular models for the circadian clock. I. The chronon concept. *J. theor. Biol.* **15,** 240.
Eilenberg, S. and Steenrod, N. (1952). "Foundations of Algebraic Topology", Princeton University Press.
Elgjo, K. (1973). Epidermal chalone: cell cycle specificity of two epidermal growth inhibitors. *Natn. Cancer Inst. Monogr.* **38,** 71.
Elsdale, T., Pearson, M. and Whitehead, M. (1976). Abnormalities in somite segmentation following heat shock to *Xenopus* embryos. *J. Embryol. exp. Morph.* **35,** 625.
Engelman, W. (1966). Effect of light and dark pulses on the emergence rhythm of *Drosophila pseudo-obscura. Experientia* **22,** 1.
Engleberg, J. (1968). On deterministic origins of mitotic variability. *J. theor. Biol.* **20,** 249.
Faraday Society Symposium (1974) **9** Physical Chemistry of Oscillatory Phenomena.
Fowler, D. H. (1972). The Riemann–Hugoniot catastrophe and van der Waal's equation. *In* "Towards a Theoretical Biology" (ed. C. H. Waddington), Vol. 4, p. 1, Edinburgh University Press.
Frank, E. K. (1970). A mathematical model of synchronized periodic growth of cell populations. *J. theor. Biol.* **26,** 373.
Frankel, J. (1974). Positional information in unicellular organisms. *J. theor. Biol.* **47,** 439.
Frankfurt, O. S. (1971). Epidermal chalone. *Expl Cell Res.* **64,** 140.
Fraser, A. and Tiwari, J. (1974). Genetical feed-back repression. II. Cyclic genetic systems. *J. theor. Biol.* **47,** 397.
Freed, J. J. and Schatz, S. A. (1969). Chromosome aberrations in cultured cells deprived of single essential amino acids. *Expl Cell Res.* **55,** 393.
French, V., Bryant, P. J. and Bryant, S. V. (1976). Specification of position in an epimorphic field. *Science, N.Y.* (in the press).
French, V. and Bullière, D. (1975a). Nouvelles données sur la détermination de la position des cellules épidermique sur un appendice de blatte. *C.r. hebd. Séanc. Acad. Sci., Paris* **218,** 53.
French, V. and Bullière, D. (1975b). Etudes sur la détermination de la position des cellules épidermiques; ordonnencement des cellules autour d'un appendice de blatte; démonstrations du concept de génératrice. *C.r. hebd. Séanc. Acad. Sci., Paris* **218,** 295.
Frye, L. D. and Edidin, M. (1970). The rapid mixing of cell surface antigens after formation of mouse and human heterokaryons. *J. Cell Sci.* **7,** 319.
Gallant, J., Margason, G. and Finch, B. (1972). On the turnover of ppGpp in *Escherichia coli. J. biol. Chem.* **247,** 6055.
Garcia-Giralt, E., Lasalvia, E., Florentin, I. and Mathé, G. (1970). Evidence for a lymphocyte chalone. *Eur. J. clin. Biol. Res.* **15,** 1012.

Gaze, R. M., Jacobson, M. and Szekely, G. (1963). The retino-tectal projection in *Xenopus* with compound eyes. *J. Physiol., Lond.* **165,** 484.
Gaze, R. M., Jacobson, M. and Szekely, G. (1965). On the formation of connections by compound eyes in *Xenopus. J. Physiol., Lond.* **176,** 409.
Gaze, R. M., Keating, M. J. and Straznicky, K. (1970). The re-establishment of retino-tectal projections after uncrossing the optic chiasma in *Xenopus laevis* with one compound eye. *J. Physiol., Lond.* **207,** 51P.
Gaze, R. M. and Keating, M. J. (1972). The visual system and "Neuronal Specificity". *Nature, Lond.* **237,** 375.
Gerhardt, J. C. and Pardee, A. B. (1963). The effect of the feed-back inhibitor, CTP, on subunit interaction in aspartate transcarbamylase. *Cold Spring Harb. Symp. quant. Biol.* **28,** 491.
Gerhardt, J. C. and Schachman, H. K. (1965). Distinct subunits for the regulation and catalytic activity of aspartate transcarbamylase. *Biochemistry, N.Y.* **4,** 1054.
Gerisch, G. (1968). Cell aggregation and differentiation in *Dictoyostelium. Curr. Top. devl Biol.* **3,** 157.
Gierer, A. and Meinhardt, H. (1972). A theory of biological pattern formation. *Kybernetik* **12,** 20.
Gilbert, W. and Müller-Hill, B. (1966). Isolation of the *lac* repressor. *Proc. natn. Acad. Sci. U.S.A.* **56,** 1891.
Gitler, C. (1972). Plasticity of biological membranes. *A. Rev. Biophys. Bioeng.* **1,** 1.
Glansdorff, P. and Prigogine, I. (1971). "Thermodynamics of Structure, Stability, and Fluctuations", Wiley Interscience, New York.
Glass, L. (1973). Instability and mitotic patterns in tissue growth. *J. Dynamic Systems, Measure and Control* **95G,** 324.
Glass, L. and Kauffman, S. A. (1973). The logical analysis of continuous, non-linear biochemical networks. *J. theor. Biol.* **39,** 103.
Goldbeter, A. (1975). Mechanism of oscillatory synthesis of cyclic AMP in *Dictyostelium discoideum. Nature, Lond.* **253,** 540.
Goodwin, B. C. (1963). "Temporal Organization in Cells", Academic Press, London and New York.
Goodwin, B. C. (1968). The division of cells and the fusion of ideas. *In* "Towards a Theoretical Biology" (ed. C. H. Waddington), Vol. 1, p. 134, Edinburgh University Press.
Goodwin, B. C. (1969a). Synchronization of *E. coli* B in chemostat by periodic phosphate feeding. *Eur. J. Biochem.* **10,** 511.
Goodwin, B. C. (1969b). Control dynamics of β-galactosidase in relation to the bacterical cell cycle. *Eur. J. Biochem.* **10,** 515.
Goodwin, B. C. (1969c). Growth dynamics and synchronization of cells. *Symp. Soc. gen. Microbiol.* **19,** 223.
Goodwin, B. C. (1970a). Model of the bacterial growth cycle: statistical dynamics of a system with asymptotic orbital stability. *J. theor. Biol.* **28,** 375.
Goodwin, B. C. (1970b). Biological stability. *In* "Towards a Theoretical Biology" (ed. C. H. Waddington), Vol. 3, p. 1, Edinburgh University Press.
Goodwin, B. C. (1972a). An analysis of the retino-tectal projection of the amphibian visual system. *In* "Lectures in Mathematics in the Life Sciences" (ed. J. D. Cowan), p. 75, American Mathematical Society, Providence, Rhode Island.
Goodwin, B. C. (1972b). Biology and meaning. *In* "Towards a Theoretical Biology" (ed. C. H. Waddington), Vol. 4, p. 259, Edinburgh University Press.
Goodwin, B. C. (1973). Embryogenesis and cognition. *In* "Cybernetics and Bionics" (eds W. D. Keidel, W. Händler and M. Spreng), p. 47, Oldenbourg-Verlag, Munich.
Goodwin, B. C. (1974). Excitability and spatial order in membranes of developing systems. *Faraday Soc. Symp.* **9,** 226.
Goodwin, B. C. (1975). A membrane model for polar ordering and gradient formation. *Adv. chem. Phys.* **29,** 269.
Goodwin, B. C. and Cohen, M. H. (1969). A phase-shift model for the spatial and temporal organization of developing systems. *J. theor. Biol.* **25,** 49.

Gorini, L., Gundersen, W. and Burger, H. (1961). Genetics of regulation of enzyme synthesis in the arginine biosynthetic pathway of *Escherichia coli*. *Cold Spring Harb. Symp. quant. Biol.* **26,** 173.

Gorini, L. and Kalman, S. M. (1963). Control by uracil of carbamyl phosphate synthesis in *Escherichia coli*. *Biochim, biophys. Acta* **69,** 355.

Graham, R. (1973). Statistical theory of instabilities in stationary non-equilibrium systems with applications to lasers and non-linear optics. *In* "Quantum Statistics in Optics and Solid-State Physics" (ed. G. Höhler), p. 1, Springer-Verlag, Berlin.

Hadorn, E. (1966). Konstanz, Wechsel, und Typus der Determination und Differenzierung in Zeller aus männlichen Genitalanlager von *Drosophila melanogaster* nach Dauerkultur *in vivo*. *Devl Biol.* **13,** 424.

Hagins, W. A. (1972). The visual process: excitatory mechanisms in the primary receptor cells. *A. Rev. Biophys. Bioeng.* **1,** 131.

Haken, H. (1973). Synergetics—towards a new discipline. *In* "Co-operative Phenomena" (eds H. Haken and M. Wagner), p. 363, Springer-Verlag, Berlin.

Halberg, F. (1963). Circadian (about 24 h) rhythms in experimental medicine. *Proc. R. Soc. Med.* **56,** 253.

Hämmerling, J. (1934). Uber Formbildende Substanzen bei *Acetabularia mediterranea*, ihre Räumliche und Zeitliche Verteilung und ihre Herkunft. *Wilhelm Roux Arch. EntwMech. Org.* **131,** 1.

Hämmerling, J. (1963). Nucleo-cytoplasmic interactions in *Acetabularia* and other cells. *A. Rev. Pl. Physiol.* **14,** 65.

Hampé, A. (1959). Contribution à l'étude du développement et de la régulation des déficiences et des excédents dans la patte de l'embryon de poulet. *Archs Anat. microsc. Morph. exp.* **48,** 1345.

Harris, H. (1967). The reactivation of the red cell nucleus. *J. Cell Sci.* **2,** 23.

Harris, H., Sidebottom, E., Grace, D. M. and Bramwell, M. E. (1969). The expression of genetic information: a study with hybrid animal cells. *J. Cell Sci.* **4,** 499.

Harrison, R. G. (1935). On the origin and development of the nervous system studied by the methods of experimental embryology. *Proc. R. Soc.* B. **118,** 155.

Hartwell, L. H., Culotti, J., Pringle, J. R. and Reid, B. J. (1974). Genetic control of the cell division cycle in yeast. *Science, N.Y.* **183,** 46.

Hastings, J. W. (1960). Biochemical aspects of rhythms: phase shifting by chemicals. *Cold Spring Harb. Symp. quant. Biol.* **25,** 131.

Hazeltine, W. A. and Block, R. (1973). Synthesis of guanosine tetra- and penta-phosphate requires the presence of a codon-specific, uncharged transfer ribonucleic acid in the acceptor site of ribosomes. *Proc. natn. Acad. Sci. U.S.A.* **70,** 1564.

Hearon, J. Z. (1952). The kinetics of linear systems with special reference to periodic reactions. *Bull. math. Biophys.* **15,** 121.

Helmstetter, C. E. and Cooper, S. (1968). DNA synthesis during the division cycle of rapidly growing *Escherichia coli* B/r. *J. molec. Biol.* **31,** 507.

Helmstetter, C. E. and Pierucci, O. (1968). Cell division during inhibition of deoxyribonucleic acid synthesis in *Escherichia coli*. *J. Bact.* **45,** 1627.

Herbert, D., Ellsworth, R. and Telling, R. C. (1956). The continuous culture of bacteria: a theoretical and experimental study. *J. gen. Microbiol.* **14,** 601.

Herth, W. and Sander, K. (1973). Mode and timing of body pattern formation (regionalization) in the early embryonic development of *Cyclorrhaphic Dipterans* (*Protophormia, Drosophila*). *Wilhelm Roux Arch. EntwMech. Org.* **172,** 1.

Hertwig, R. (1908). Uber neue Probleme der Zellenlehre. *Arch. Zellforsch.* **1,** 1.

Hess, B. and Boiteux, A. (1971). Oscillatory phenomena in biochemistry. *A. Rev. Biochem.* **40,** 237.

Higgins, J. (1967). The theory of oscillating reactions. *Ind. Engng Chem.* **59,** 19.

Hirsch, H. R. and Engleberg, J. (1966). Decay of cell synchronization: solutions of the cell-growth equation. *Bull. math. Biophys.* **28,** 391.

Hitchcock, S. E. (1973). Regulation of muscle contraction. Effect of calcium on the affinity of troponin for actin and tropomyosin. *Biochemistry, N.Y.* **12,** 2509.

Hocking, J. C. and Young, G. S. (1961). "Topology", Addison-Wesley, New York.
Holtzer, H. (1963). Mitosis and cell transformation. *In* "General Physiology of Cell Specialization" (eds D. Mazia and A. Tyler), p. 80, McGraw-Hill, New York.
Hondius-Boldingh, W. and Laurence, E. B. (1968). Extraction, purification and preliminary characterisation of the epidermal chalone: a tissue specific mitotic inhibitor obtained from vertebrate skin. *Eur. J. Biochem.* **5,** 191.
Huang, K. (1963). "Statistical mechanics", Wiley, New York.
Iberall, A. S. (1969). New thoughts on biocontrol. *In* "Towards a Theoretical Biology" (ed. C. H Waddington), Vol. 2, p. 166, Edinburgh University Press.
Inoye, M. (1969). Unlinking of cell division from DNA replication in a temperature-sensitive DNA synthesis mutant of *Escherichia coli*. *J. Bact.* **99,** 842.
Iversen, O. H. (1961). The regulation of cell number in epidermis. A cybernetic point of view. *Acta path. microbiol. scand.* **148,** 91.
Iversen, O. H. (1970). Some theoretical considerations on chalones and the treatment of cancer: a review. *Cancer Res.* **30,** 1481.
Jacob, F. and Monod, J. (1961). Genetic regulatory mechanisms in the synthesis of proteins. *J. molec. Biol.* **3,** 318.
Jacob, F., Brenner, S. and Cuzin, F. (1963). On the regulation of DNA replication in bacteria. *Cold Spring Harb. Symp. quant. Biol.* **28,** 329.
Jacob, F., Ryter, A. and Cuzin, F. (1966). On the association between DNA and membrane in bacteria. *Proc. R. Soc.* B. **164,** 267.
Jacobson, C. O. (1959). The localization of the presumptive cerebral regions in the neural plate of the axolotl larvae. *J. Embryol. exp. Morph.* **7,** 1.
Jacobson, C. O. (1964). Motor nuclei, cranial nerve roots, and fibre pattern in the medulla oblongata after reversal experiments on the neural plate of axolotl larvae. *Zool. Bidr. Upps.* **36,** 73.
Jacobson, M. (1968a). Development of neuronal specificity in retinal ganglion cells of *Xenopus*. *Devl Biol.* **17,** 202.
Jacobson, M.(1968b). Cessation of DNA synthesis in retinal ganglion cells correlated with time of specification of their central connections. *Devl Biol.* **17,** 219.
Jacobson, M. (1975). Discussion. *In* "Cell Patterning" (eds R. Porter and I. Rivers), Ciba Foundation Symposium, Vol. 29, p. 333, Elsevier, Amsterdam.
Jacobson, M. and Levine, R. L. (1975). Plasticity in the adult frog brain: filling the visual scotoma after excision of translocation of parts of the optic tectum. *Brain Res.* **88,** 339.
Jacobson, M. and Levine, R. L. (1975). Stability of implanted duplicate tectal positional markers serving as targets for optic axons in adult frogs. *Brain Res., Osaka* **92,** 468.
Jaffe, L. (1966). Electrical currents through the developing *Fucus* egg. *Proc. natn. Acad. Sci. U.S.A.* **56,** 1102.
Jaffe, L. (1968). Localization in the developing egg and the general role of localizing currents. *In* "Advances in Morphogenesis" (eds M. Abercrombie and J. Brachet), Vol. 7, p. 295, Academic Press, New York and London.
Jerka-Dziadosz, M., (1964). *Urostyla cristata* sp. n. (Urostylidae, Hypotrichida): the morphology and morphogenesis. *Acta Protozool.* **2,** 120.
Jung, C. and Rothstein, A. (1967). Cation metabolism in relation to cell size in synchronously grown tissue culture cells. *J. gen. Physiol.* **50,** 917.
Kacser, H. (1957). Some physico-chemical aspects of biological organization. Appendix to "The Strategy of the Genes" (C. H. Waddington), Allen and Unwin, London.
Kaczanowska, J. (1974). The pattern of morphogenetic control in *Chilodonella cucullulus*. *J. exp. Zool.* **187,** 47.
Karakashian, M. W. and Hastings, J. W. (1962). The inhibition of a biological clock by actinomycin D. *Proc. natn. Acad. Sci. U.S.A.* **48,** 2130.
Kauffman, S. A. (1969). Metabolic stability and epigenesis in randomly constructed genetic nets. *J. theor. Biol.* **22,** 437.
Kauffman, S. A. (1973). Control circuits for determination and transdetermination. *Science, N.Y.* **181,** 310.

Kauffman, S. A. (1974). 'Measuring a mitotic oscillator: the arc discontinuity. *Bull. math. Biol.* **36,** 171.

Kauffman, S. A. and Wille, J. J. (1975). The mitotic oscillator in *Physarum polycaphalum*. *J. theor. Biol.* **55,** 47.

Keller, E. F. and Segel, L. A. (1970). Initiation of slime mold aggregation viewed as an instability. *J. theor. Biol.* **26,** 399.

Kendall, D. G. (1948). On the role of variable generation time in the development of a stochastic birth process. *Biometrika* **35,** 316.

Kieny, M. (1964). Etude du mécanisme de la régulation dans le développement du bourgeon de membre de l'embryon de poulet. *Devl Biol.* **9,** 197.

Kishimoto, S. and Lieberman, I. (1964). Synthesis of RNA and protein required for the mitosis of mammalian cells. *Expl Cell Res.* **36,** 92.

Kivilaakso, E. and Rytomäa, T. (1971). Erythrocyte chalone, a tissue-specific inhibitor of cell proliferation in the erythron. *Cell Tiss. Kinet.* **4,** 1.

Konijn, T. M., Barkley, D. S., Chang, Y.-Y. and Bonner, J. T. (1968). Cyclic AMP: a naturally occurring acrasin in the cellular slime molds. *Am. Nat.* **102,** 225.

Korn, A. P., Henkelman, R. M., Ottensmeyer, F. P. and Till, J. E. (1973). Investigations of a stochastic model of haemopoiesis. *Expl Hemat.* **1,** 362.

Kornacker, K. (1969a). Cognitive processes in physics and physiology. *In* "Towards a Theoretical Biology" (ed. C. H. Waddington), Vol. 2, p. 248, Edinburgh University Press.

Kornacker, K. (1969b). Physical principles of active transport and electrical excitability. *In* "Biological Membranes" (ed. R. M. Dowben), p. 39, Little and Brown, Boston.

Kornacker, K. (1972). Living aggregates of non-living parts: a generalized statistical mechanical theory. *Prog. theor. Biol.* **2,** 1.

Koshland, D. E. (1970). The molecular basis for enzyme regulation. *In* "The Enzymes" (ed. P. D. Boyer), Vol. 1, p. 342, Academic Press, London and New York.

Kubitschek, H. (1966). Generation times: ancestral dependence and dependence upon cell size. *Expl Cell Res.* **43,** 30.

Lawrence, P. A. (1966). Gradients in the insect segment: the orientation of hairs in the milkweed bug *Oncopeltus fasciatus*. *J. exp. Biol.* **44,** 607.

Lawrence, P. A. (1970). Polarity and patterns in the postembryonic development of insects. *Adv. Insect Physiol.* **7,** 197.

Lawrence, P. A. (1974). Cell movement during pattern regulation in *Oncopeltus*. *Nature, Lond.* **248,** 609.

Lawrence, P. A. (1975). The structure and properties of a compartment border: the intersegmental boundary in *Oncopeltus*. *In* "Cell Patterning" (eds R. Porter and J. Rivers), Ciba Foundation Symposium, Vol. 29, p. 3, Elsevier, Amsterdam.

Lawrence, P. A., Crick, F. H. C. and Munro, M. (1972). 'A gradient of positional information in an insect, *Rhodnius*. *J. Cell Sci.* **11,** 815.

Lederberg, J. and Tatum, E. L. (1946). Novel genotypes in mixed cultures of biochemical mutants of bacteria. *Cold Spring Harb. Symp. quant. Biol.* **11,** 113.

Lentz, T. H. (1965). *Hydra*: induction of supernumerary heads by isolated neurosecretory granules. *Science, N.Y.* **150,** 633.

Levine, R. L. and Jacobson, M. (1974). Deployment of optic nerve fibres is determined by positional markers in the frog's tectum. *Expl Neurol.* **43,** 527.

Le Witt, M. A. J. (1972). "Synchronisation of mammalian cells in continuous culture". Ph.D. thesis, Sussex University.

Locke, M. (1959). The cuticular pattern in an insect, *Rhodnius prolixus*, Stal. *J. exp. Biol.* **36,** 459.

Locke, M. (1966). The cuticular pattern in an insect. The behaviour of grafts in segmented appendages. *J. Insect Physiol.* **12,** 397.

Loomis, W. F. and Magasanik, B. (1966). Nature of the effector of catabolite repression of β-galactosidase in *E. coli*. *J. Bact.* **92,** 170.

Lowenstein, W. R. (1967). On the genesis of cellular communication. *Devl Biol.* **15,** 503.

Lowenstein, W. R. and Kanno, Y. (1967). Intercellular communication and tissue growth. *J. Cell Biol.* **33,** 225.

Lund, E. J. (1921). Experimental control of organic polarity by the electrical current. I. Effects of electrical currents on regenerating internodes of *Obelia commisuralis*. *J. exp. Zool.* **34,** 471.

Maaløe, O. and Kjeldegaard, N. O. (1966). "Control of Macromolecular Synthesis", W. A. Benjamin, New York.

Maaløe, O. Kurland, C. G. (1963). The integration of protein and ribonucleic acid synthesis in bacteria. *In* "Cell Growth and Cell Division" (ed. R. J. C. Harris), p. 93, Academic Press, London and New York.

Maas, W. K. (1961). Studies in repression of arginine biosynthesis in *Escheria coli*. *Cold Spring Harb. Symp. quant. Biol.* **26,** 183.

Makman, R. S. and Sutherland, E. Q. (1965). Adenosine 3′,5′-phosphate in *Escherichia coli*. *J. biol. Chem.* **240,** 1309.

Marcus, W. (1962). Untersuchungen über die Polarität der Rumphaut von Schmetterlingen. *Wilhelm Roux Arch. EntwMech. Org.* **154,** 56.

Markowitz, D. (1971). Collective modes and wave-like solutions to cellular control equations. *J. theor. Biol.* **31,** 475.

Markowitz, D. and Nisbet, R. M. (1973). Co-operativity in biological machines. *J. theor. Biol.* **39,** 653.

Marks, F. (1973a). A tissue-specific factor inhibiting DNA synthesis in mouse epidermis. *Natn. Cancer Inst. Monogr.* **38,** 79.

Marks, F. (1973b). The second messenger system of mouse epidermis. III. Guanyl cyclase. *Biochim. biophys. Acta* **309,** 349.

Marr, A. G., Painter, P. R. and Nilson, E. H. (1959). Growth and division of individual bacteria. *Symp. Soc. gen. Microbiol.* **19,** 237.

Martinez, H. H. (1972). Morphogenesis and chemical dissipative structures in a computer-simulated case study. *J. theor. Biol.* **36,** 479.

Masters, M. and Broda, P. (1971). Evidence for the bidirectional replication of the *E. coli* chromosome. *Nature, New Biology* **232,** 137.

Mayersbach von, H. (ed.) (1969). "Cellular Aspects of Biorhythm", Springer-Verlag, Berlin.

Maynard Smith, J. (1960). Continuous, quantized and model variation. *Proc. R. Soc.* B. **152,** 397.

McClare, C. W. F. (1971). Chemical machines, Maxwell's Daemon and living organisms. *J. theor. Biol.* **30,** 1.

McClare, C. W. F. (1972). A "Molecular Energy" muscle model. *J. theor. Biol.* **35,** 569.

McLane, S. (1971). "Categories for the Working Mathematician", Springer, New York.

McMahon, D. (1973). A cell-contact model for cellular position determination in development. *Proc. natn. Acad. Sci. U.S.A.* **70,** 2396.

Meinhardt, H. and Gierer, A. (1974). Applications of a theory of biological pattern formation based upon lateral inhibition. *J. Cell Sci.* **15,** 321.

Mohr, U., Althoff, J. Kinzel, U., Süss, R. and Volm, M. (1968). Melanoma regression induced by 'chalone': a new tumour inhibiting principle acting *in vivo*. *Nature, Lond.* **220,** 138.

Monod, J. (1947). The phenomenon of enzymatic adaptation and its bearings on problems of genetics and cellular differentiation. *Growth* **11,** 223.

Monod, J. (1950). La technique de culture continue: théorie et applications. *Annls Inst. Pasteur Lille* **79,** 370.

Monod, J. and Jacob, F. (1961). Teleonomic mechanisms in cellular metabolism, growth and differentiation. *Cold Spring Harb. Symp. quant. Biol.* **26,** 389.

Monod, J., Wyman, J. and Changeux, J.-P. (1965). On the nature of allosteric transitions: a plausible model. *J. molec. Biol.* **12,** 88.

Moser, H. (1958). "The dynamics of bacterial populations maintained in the chemostat", Publication No. 614, Carnegie Institution of Washington.

Nanjundiah, V. (1974). A differential chemotactic response of slime mould amoebae to regions of the early amphibian embryo. *Expl Cell Res.* **86,** 408.

Nanney, D. L. (1968). Cortical patterns in morphogenesis. *Science, N.Y.* **160,** 496.

Naora, H. (1973). Nuclear RNA. *In* "The Ribonucleic Acids" (eds P. R. Stewart and D. S. Latham), p. 37, Springer-Verlag, New York.

Nelbach, M. E., Pigiet, V. P., Gerhardt, J. C. and Schachman, H. K. (1972). A role for zinc in the quaternary structure of aspartate transcarbamylase from *Escherichia coli*. *Biochemistry, N.Y.* **11,** 315,

Newman, S. A. (1972). A source of stability in metabolic networks. *J. theor. Biol.* **35,** 227.

Newman, S. A. (1974). The interaction of the organizing regions in *Hydra* and its possible relation to the role of the cut end in regeneration. *J. Embryol. exp. Morph.* **31,** 541.

Nicolson, G. L. and Singer, S. J. (1971). Ferrition in conjugated plant agglutinins as specific saccharide stains for electron microscopy: application to saccharides bound to cell membranes. *Proc. natn. Acad. Sci. U.S.A.* **68,** 942.

Njus, D., Sulsman, F. M. and Hastings, J. W. (1974). Membrane model for the circadian clock. *Nature, Lond.* **248,** 116.

Novak, B. and Bentrup, F. W. (1972). An electrophysiological study of regeneration in *Acetabularia mediterranea*. *Planta* **108,** 227.

Novick, A. and Szilard, L. (1950). Experiments with the chemostat on spontaneous mutations of bacteria. *Proc. natn. Acad. Sci. U.S.A.* **36,** 708.

Novick, A. and Szilard, L. (1954). Experiments with the chemostat on the rates of amino acid synthesis in bacteria. *In* "Dynamics of Growth Processes", p. 21, Princeton University Press.

Nübler-Jung, K. (1974). Cell migration during pattern reconstitution in the insect segment (*Dysdercus intermedius* Dist., Heteroptera). *Nature, Lond.* **248,** 610.

Nucitelli, R. and Jaffe, L. F. (1974). Spontaneous current pulses through developing fucoid eggs. *Proc. natn. Acad. Sci. U.S.A.* **71,** 4855.

Pardee, A. B. (1968). Control of cell division: models from micro-organisms. *Cancer Res.* **28,** 1802.

Passano, L. M. and McCullough, C. B. (1965). Coordinating systems and behaviour in *Hydra*. *J. exp. Biol.* **42,** 205.

Pastan, I. and Perlman, R. (1970). Cyclic adenosine monophosphate in bacteria. *Science, N.Y.* **169,** 339.

Pattee, H. H. (1974). Discrete and continuous processes in computers and brains. *In* "Lecture Notes in Biomathematics. IV. Physics and mathematics of the nervous system". (eds M. Conrad, W. Güttinger, and M. Dal Cin), p. 128, Springer-Verlag, Berlin.

Paul, J. (1972). General theory of chromosome structure and gene activation in eukaryotes. *Nature, Lond.* **238,** 444.

Pederson, F. S., Lane, E. and Kalgora, N. O. (1973). Codon specific, tRNA dependent *in vitro* synthesis of ppGpp and pppGpp. *Nature, New Biology* **243,** 13.

Peterkovsky, A. and Gazdor, C. (1974). Glucose inhibition of adenylate cyclase in intact cells of *Escherichia coli* B. *Proc. natn. Acad. Sci. U.S.A.* **71,** 2324.

Piepho, H. (1955). Über die polare Orientierung der Bälge und Schappen auf dem Schmatterlings-rumpf. *Biol. Zbl.* **74,** 467.

Piérard, A., Glansdorff, N., Mergeay, M. and Wiame, J. M. (1965). Control of the biosynthesis of carbamyl phosphate in *Escherichia coli*. *J. molec. Biol.* **14,** 23.

Pilgrim, C. (1967). Autoradiographic investigations with ^{3}H-thymidine on the influence of the diurnal rhythm in cell proliferation kinetics. *In* "The cellular aspects of biothythms" (ed. H. von Mayersbach), p. 100, Springer-Verlag, Berlin.

Pilgrim, C., Erb, W. and Maurer, W. (1963). Diurnal fluctuations in the numbers of DNA synthesizing nuclei in various mouse tissues. *Nature, Lond.* **199,** 863.

Pittendrigh, C. S. (1960). Circadian rhythms and the circadian organization of living systems. *Cold Spring Harb. Symp. quant. Biol.* **25,** 159.

Pittendrigh, C. S. (1965). On the mechanism of the entrainment of a circadian rhythm by light cycles. *In* "Circadian Clocks" (ed. J. Aschoff), p. 277, North-Holland, Amsterdam.

Pittendrigh, C. S. (1966). The circadian oscillation in *Drosophila pseudo-obscura* pupae: a model for the photoperiodic clock. *Z. Pflanzenphysiol.* **54,** 275.

Pittendrigh, C. S. and Bruce, V. G. (1957). An oscillator model for biological clocks. *In* "Rhythmic and Synthetic Processes in Growth" (ed. D. Rudnick), p. 75, Princeton University Press.

Porcellati, G. and di Jeso, F. (eds) (1971). Membrane-bound enzymes. *Adv. exp. Biol. Med.* **14,** 71.

Potten, C. S. (1974). The epidermal proliferative units: the possible role of the central basal cell. *Cell Tiss. Kinet.* **7,** 77.

Powell, E. and Errington, F. (1963). Generation time of individual bacteria: some corroborative measurements. *J. gen. Microbiol.* **31,** 315.

Prescott, D. M. (1956). Relation between cell growth and cell division. *Expl Cell Res.* **11,** 86.

Prescott, D. M. and Goldstein, L. (1967). Nuclear-cytoplasmic interaction in DNA synthesis. *Science, N.Y.* **155,** 469.

Prestige, M. C. and Willshaw, D. J. (1975). On a role for competition in the formation of patterned neural connexions. *Proc. R. Soc.* B. **190,** 77.

Prevost, C. and Moses, V. (1967). Pool sizes of metabolic intermediates and their relation to glucose repression of β-galactosidase synthesis in *E. coli. Biochem. J.* **103,** 349.

Prigogine, I. (1969). Structure, dissipation and life. *In* "Theoretical Physics and Biology" (ed. M. Marois, p. 23, North-Holland, Amsterdam.

Pritchard, R. H., Barth, P. T. and Collins, J. (1969). Control of DNA synthesis in bacteria. *Symp. Soc. gen. Microbiol.* **19,** 263.

Prothero, J. W. and Tyler, R. W. (1975). A model of thymocyte proliferation. *J. theor. Biol.* **51,** 357.

Rahn, O. (1932). A chemical explanation of the variability of the growth rate. *J. gen. Physiol.* **15,** 257.

Rapp, P. E. (1975). A theoretical investigation of a large class of biochemical oscillators. *Math. Biosciences* **25,** 165.

Reanney, D. C. (1975). A regulatory role for viral RNA in eukaryotes. *J. theor. Biol.* **49,** 461.

Richter, C. P. (1965). "Biological Clocks in Medicine and Psychiatry", Charles C. Thomas, Springfield, Illinois.

Roberts, R. B., Abelson, P. H., Cowie, D. B., Bolton, E. T. and Britten, R. J. (1955). "Studies of biosynthesis in *Escherichia coli*", Publication No. 607, Carnegie Institution of Washington.

Robertson, A., Drage, D. J. and Cohen, M. H. (1972). Control of aggregation in *Dictyostelium discoideum* by an externally applied periodic pulse of cyclic AMP. *Science, N.Y.* **175,** 333.

Rose, S. M. (1963). Polarized control of regional structure in *Tubularia. Devl Biol.* **7,** 488.

Rosenbusch, J. B. and Weber, K. (1971). Subunit structure of aspartate transcarbamylase from *Escherichia coli. J. biol. Chem.* **246,** 1644.

Rubin, J. and Robertson, A. (1975). The tip of the *Dictyostelium discoideum* pseudoplasmodium as an organizer. *J. Embryol. exp. Morph.* **33,** 227.

Rusch, H. P. (1970). Some biochemical events in the life cycle of *Physarum polycephalum. Adv. Cell Biol.* **1,** 297.

Rytomäa, T. and Kiviniemi, K. (1968a). Control of cell production in rat chloroleukaemia by means of the granulocytic chalone. *Nature, Lond.* **220,** 136.

Rytomäa, T. and Kiviniemi, K. (1968b). Control of granulocyte production. *Cell Tiss. Kinet.* **1,** 329.

Sachsenmaier, W., Remy, U. and Plattner-Schobel, R. (1972). Initiation of synchronous mitosis in *Physarum polycephalum. Expl Cell Res.* **73,** 41.

Saetren, H. A. (1956). A principle of auto-regulation of growth. Production of organ specific mitosis-inhibitor in kidney. *Expl. Cell Res.* **11,** 229.

Sandahchiev, L. S., Puchkova, L. I. and Pikalov, A. V. (1972). Subcellular localization of morphogenetic factors in anucleate *Acetabularia* at the stages of genetic information transfer and expression. *In* "Biology and Radiobiology of Anucleate systems. II. Plant Cells" (eds S. Bonotto, R. Goutier, R. Kirchman and J.-R. Maisin), p. 297, Academic Press, New York and London.

Sander, G. and Pardee, A. B. (1972). Transport changes in synchronously growing CHO and L cells. *J. cell. Physiol.* **80,** 267.

Saunders, J. W. (1948). The proximo-distal sequence of origin of the parts of the chick wing and the role of the ectoderm. *J. exp. Zool.* **108,** 363.

Saunders, J. W. and Gosseling, M. T. (1968). Ectodermal-mesodermal interactions in the origin of limb symmetry. *In* "Epithelial-Mesenchymal Interactions" (eds R. Fleischmajer and R. E. Billingham), p. 78, Williams and Wilkins, Baltimore.
Savageau, M. A. (1974). Genetic regulatory mechanisms and the ecological niche of *Escherichia coli*. *Proc. natn. Acad. Sci. U.S.A.* **71,** 2453.
Schaechter, M., Williamson, J. P., Houd, J. R. and Koch, A. L. (1962). Growth cell and nuclear divisions in some bacteria. *J. gen. Microbiol.* **29,** 421.
Schaller, C. H. (1973). Isolation and characterization of low molecular weight substances activating head and bud formation in *Hydra*. *J. Embryol. exp. Morph.* **29,** 27.
Schrödinger, I. W. (1948). "What Is Life?" Cambridge University Press.
Schubiger, G. (1971). Regeneration, duplication and transdetermination in fragments of the leg discs of *Drosophila melanogaster*. *Devl Biol.* **26,** 277.
Schweiger, E., Wallraff, H. G. and Schweiger, H. G. (1964). Endogenous circadian rhythms in cytoplasm of *Acetabularia*; influence of the nucleus. *Science, N.Y.* **146,** 658.
Segel, L. A. (1972). On collective motions of chemotactic cells. *In* "Lectures on Mathematics in the Life Sciences" (ed. J. D. Cowan), Vol. 4, p. 1, American Mathematical Society, Providence, Rhode Island.
Shaffer, B. M. (1962). The Acrasina. *In* "Advances in Morphogenesis" (eds M. Abercrombie and J. Brachet), Vol. 2, p. 109, Academic Press, New York and London.
Siekevitz, P. (1972). Biological membranes: the dynamics of their organization. *A. Rev. Physiol.* **34,** 117.
Singer, S. J. and Nicolson, G. L. (1972). The fluid mosaic model of the structure of cell membranes. *Science, N.Y.* **175,** 720.
Smith, J. A. and Martin, L. (1973). Do cells cycle? *Proc. natn. Acad. Sci. U.S.A.* **70,** 1263.
Sompayrac, L. and Maaløe, O. (1973). Autorepressor model for control of DNA replication. *Nature, New Biology* **241,** 133.
Sonneborn, T. M. (1963). Does preformed cell structure play an essential role in cell heredity?. *In* "The Nature of Biological Diversity" (ed. J. M. Allen), p. 166, McGraw-Hill, Maidenhead, Berks.
Sperry, R. W. (1944). Optic nerve regeneration with return of vision in anurans. *J. Neurophysiol.* **7,** 57.
Sperry, R. W. (1951). Regulative factors in the orderly growth of neural circuits. *Growth* **15,** 63.
Sperry, R. W. (1955). Problems in the biochemical specification of neurons. *In* "Biochemistry of the Developing Nervous System" (ed. H. Waelsch), p. 74, Academic Press, New York and London.
Sperry, R. W. (1963). Chemoaffinity in the orderly growth of nerve fibre patterns and connections. *Proc. natn. Acad. Sci. U.S.A.* **50,** 703.
Spiegelman, S. (1948). Differentiation as the controlled production of unique enzymatic patterns. *Symp. Soc. exp. Biol.* **2,** 286.
Spiegelman, S. and Reiner, J. (1947). The formation and stabilization of an adaptive enzyme in the absence of its substrate. *J. gen. Physiol.* **31,** 175.
Stanley, H. E. (1971). "Introduction to Phase Transitions and Critical Phenomena", Oxford University Press.
Stent, G. S. and Brenner, S. (1961). A genetic locus for the regulation of ribonucleic acid synthesis. *Proc. natn. Acad. Sci. U.S.A.* **47,** 2005.
Stockdale, F. E. and Topper, Y. J. (1966). The role of DNA synthesis and mitosis in hormone-dependent differentiation. *Proc. natn. Acad. Sci. U.S.A.* **56,** 1283.
Stone, L. S. (1948). Functional polarization in developing and regenerating retinae of transplanted eyes. *Ann. N.Y. Acad. Sci.* **49,** 856.
Straznicky, K., Gaze, R. M. and Keating, M. J. (1974). The retino-tectal projection from a double-ventral compound eye in *Xenopus laevis*. *J. Embryol. exp. Morph.* **31,** 123.
Stumpf, H. F. (1968). Further studies on gradient-dependent diversification in the pupal cuticle of *Galleria mellonella*. *J. Exp. Biol.* **49,** 49.
Sugita, M. (1963). The idea of a molecular automation. *J. theor. Biol.* **4,** 179.
Summerbell, D. and Lewis, J. H. (1975). Time, place and positional value in the chick wing bud. *J. Embryol. exp. Morph.* **33,** 621.

Summerbell, D., Lewis, J. H. and Wolpert, L. (1973). Positional information in chick limb morphogenesis. *Nature, Lond.* **244,** 492.

Sweeney, B. M. (1974a). Potassium content of *Gonyaulax polyedra* and phase clamps in the circadian rhythm of stimulated bioluminescence by short exposure to ethanol and valinomycin. *Pl. Physiol., Lancaster* **53,** 337.

Sweeney, B. M. (1974b). A physiological model for circadian rhythms derived from the *Acetabularia* rhythm paradoxes. *Int. J. Chronobiol.* **2,** 25.

Sweeney, B. M. and Hastings, J. W. (1958). Rhythmic cell division in populations of *Gonyaulax polyedra. J. Protozool.* **5,** 217.

Sweeney, B. M. and Haxa, E. T. (1961). Persistence of a photosynthetic rhythm in enucleated *Acetabularia. Science, N.Y.* **134,** 1361.

Sweeney, B. M., Tuffli, C. F. and Rubin, R. H. (1967). The circadian rhythm in photosynthesis in *Acetabularia* in the presence of actinomycin D, puromycin and chloramphenicol. *J. gen. Physiol.* **50,** 647.

Szekely, G. (1957). Regulations tendenzen in der Ausbildung der "funktionellen Spezifität" der Retinaanlage bei *Triturus vulgaris. Arch. EntwMech. Org.* **150,** 48.

Tartar, V. (1967). Morphogenesis in protozoa. *In* "Research on Protozoology" (ed. T.-T. Chen), Vol. 2, p. 1, Pergamon Press, Oxford.

Thom, R. (1970a). Phase transitions as catastrophes. *In* "Statistical Mechanics: New Concepts, New Problems, New Applications" (eds S. A. Rice, K. F. Freed and J. C. Light), p. 93, University of Chicago Press.

Thom, R. (1970b). Topological models in biology. *In* "Towards a Theoretical Biology" (ed. C. H. Waddington), Vol. 3, p. 89, Edinburgh University Press.

Thompson, J. M. T. (1965). Experiments in catastrophe. *Nature, Lond.* **254,** 392.

Thornley, J. H. M. (1975). Phyllotaxis. I. A mechanistic model. *Ann. Bot.* **39,** 491.

Thornley, J. H. M. (1975). Phyllotaxis. II. A description in terms of interacting logarithmic spirals. *Ann. Bot.* **39,** 509.

Tickle, C., Summerbell, D. and Wolpert, L. (1975). Positional signalling and specification of digits in chick limb morphogenesis. *Nature, Lond.* **254,** 199.

Tiwari, J. and Fraser, A. (1973). Genetic regulation by feedback repression. *J. theor. Biol.* **39,** 679.

Turing, A. M. (1952). The chemical basis of morphogenesis. *Phil. Trans. R. Soc. Ser. B.* **237,** 37.

Tyson, J. J. and Light, J. C. (1973). Properties of two-component bimolecular and trimolecular chemical reaction systems. *J. chem. Phys.* **59,** 4164.

Umbarger, H. E. (1956). Evidence for a negative feedback mechanism in the biosynthesis of isoleucine. *Science, N.Y.* **123,** 848.

Viniegra-Gonzalez, G. and Martinez, H. M. (1969). Stability of biochemical feedback systems. *Biophys. Soc. Abstracts* **13,** 210.

Vogel, H. J. (1961). Aspects of repression in the regulation of enzyme synthesis: pathway-wide control and enzyme-specific response. *Cold Spring Harb. Symp. quant. Biol.* **26,** 163.

Vonderhaar, B. K. and Topper, Y. J. (1974). A role of the cell cycle in hormone-dependent differentiation. *J. Cell Biol.* **63,** 707.

Waddington, C. H. (1972). Form and information; Epilogue. *In* "Towards a Theoretical Biology" (ed. C. H. Waddington), Vol. 4, pp. 109 and 283, Edinburgh University Press.

Walker, J. R. and Pardee, A. B. (1968). Evidence for a relationship between DNA metabolism and septum formation in *Escherichia coli. J. Bact.* **95,** 123.

Walter, C. F. (1969). Oscillations in controlled biochemical systems. *Biophys. J.* **9,** 863.

Walter, C. F. (1970). The occurrence and the significance of limit cycle behaviour in controlled biochemical systems. *J. theor. Biol.* **27,** 259.

Webster, G. C. (1971). Morphogenesis and pattern formation in hydroids. *Biol. Rev.* **46,** 1.

Weiss, P. and Kavanau, J. L. (1957). A model of growth and growth control in mathematical terms. *J. gen. Physiol.* **41,** 1.

Whitaker, D. M. (1938). The effect of hydrogen ion concentration on the induction of polarity in *Fucus* eggs. III. Gradient in hydrogen ion concentration. *J. gen. Physiol.* **21,** 833.

Wigglesworth, V. B. (1959). "The Control of Growth and Form: A Study of the Epidermal Cell in the Insect", Cornell University Press.

Wilby, O. K. and Webster, G. C. (1970). Experimental studies on axial polarity in *Hydra*. *J. Embryol. exp. Morph.* **24,** 595.

Winfree, A. T. (1967). Biological rhythms and the behaviour of populations of coupled oscillators. *J. theor. Biol.* **16,** 15.

Winfree, A. T. (1970). The temporal morphology of a biological clock. *In* "Lectures in Mathematics in the Life Sciences" (ed. M. Gerstenhaber), Vol. 2, p. 109, American Mathematical Society, Providence, Rhode Island.

Winfree, A. T. (1971). Corkscrews and singularities in fruit flies: resetting behaviour of the circadian eclosion rhythm. *In* "Biochronometry", p. 81, National Academy of Sciences, Washington.

Winfree, A. T. (1973). Resetting the amplitude of *Drosophilia*'s circadian chronometer. *J. comp. Physiol.* **85,** 105.

Winfree, A. T. (1975). Unclocklike behaviour of biological clocks. *Nature, Lond.* **253,** 315.

Wolpert, L. (1969). Positional information and the spatial pattern of cellular differentiation. *J. theor. Biol.* **25,** 1.

Wolpert, L. (1972). The concept of positional information and pattern formation. *In* "Towards a Theoretical Biology" (ed. C. H. Waddington), Vol. 4, p. 83, Edinburgh University Press.

Wolpert, L. Hicklin, J. and Hornbruch, A. (1971). Positional information and pattern regulation in regeneration of *Hydra*. *Symp. Soc. exp. Biol.* **25,** 391.

Wolpert, L., Lewis, J. and Summerbell, D. (1975). Morphogenesis of the vertebrate limb. *In* "Cell Patterning" (eds R. Porter and J. Rivers), Ciba Foundation Symposium, Vol. 29, p. 95, Elsevier, Amsterdam.

Woolley, W. H. and De Rocco, A. G. (1973). Asynchronous divisions in cell colonies. I. General considerations and a linear control model. *J. theor. Biol.* **39,** 73.

Yamazaki, I., Yokota, K. and Nakajuma, R. (1967). Analysis of the conditions causing the oscillatory oxidation of reduced nicotinamide-adenine dinucleotide by horseradish peroxidase. *Biochim. biophys. Acta* **132,** 310.

Yang, H.-L., Zubay, G., Urm, E., Reiness, G. and Cashel, M. (1974). Effects of guanosine tetraphosphate, guanosine pentaphosphate and β-γ-methylenyl-guanosine pentaphosphate on gene expression of *Escherichia coli in vitro*. *Proc. natn. Acad. Sci. U.S.A.* **71,** 63.

Yates, R. A. and Pardee, A. B. (1956). Control of pyrimidine biosynthesis in *Escherichia coli* by a feed-back mechanism. *J. biol. Chem.* **221,** 757.

Yoon, M. G. (1973). Retention of the original topographic polarity by the 180° rotated tectal reimplant in young adult goldfish. *J. Physiol., Lond.* **233,** 575.

Yoshikawa, H. (1967). 'The initiation of DNA replication in *Bacillus subtilis*. *Proc. natn. Acad. Sci. U.S.A.* **58,** 312.

Zeeman, E. C. (1972). Differential equations for the heart beat and nerve impulse. *In* "Towards a Theoretical Biology" (ed. C. H. Waddington), Vol. 4, p. 8, Edinburgh University Press.

Zeeman, E. C. (1974). Primary and secondary waves in developmental biology. *In* "Lectures on Mathematics in the Life Sciences", Vol. 7, p. 69, American Mathematical Society Providence, Rhode Island.

Zwilling, E. (1961). Limb morphogenesis. *In* "Advances in Morphogenesis" (eds M. Abercrombie and J. Brachet), Vol. 1, p. 301, Academic Press, New York and London.

APPENDIX

SOME basic features of the wave propagation model for morphogenetic gradient formation are given in this appendix in the form of results from computations on a particular realization of the model. This is based upon a one-dimensional analysis of a pair of differential equations of the form

$$\frac{\partial X}{\partial t} = D\frac{\partial^2 X}{\partial x^2}(x, t) + \beta Y(x, t) + M\delta(vt - x) - f(vt - v\sigma - x)X(x, t) - kX(x, t).$$

$$\frac{\partial Y}{\partial t} = -\beta Y(x, t) + f(vt - v\sigma - x)X(x, t)$$

where $X(x, t)$ is the concentration of diffusible metabolite, V (as used on p. 143), and $Y(x, t)$ is that of the complex, BV, taken to be the non-diffusing morphogen. The other parameters and functions in the equations are as follows: D is the diffusion constant of V; β is the decay constant of the morphogen, BV; $M\delta(vt - x)$ is the delta function multiplied by a constant (which together describe the propagating wave-front of the metabolic activity wave, moving at velocity v, with t as time and x as distance along the one-dimensional axis); $f\ (vt - v\sigma - x)$ is the pick-up function which propagates behind the wave-front (with the same velocity, v, after a delay time, σ); and k is the decay constant for X, describing the rate of conversion of V into W in the model. In carrying out the computation, the delta function was represented by a sharp Gaussian, describing the sudden switching on of enzyme E_1 and the production of V as the activity wave propagates. The computation was done in terms of dimensionless variables, so that D, t and x have only relative values. The wave propagates in both directions from the origin, but due to the symmetry of the result about the origin, one-half of the picture gives all the information.

In Figure A.1 is shown the shape of the morphogen curves after the propagation of a single wave, with different delay times, σ. The wave velocity was in every case 5 μms^{-1}, and the pick-up function had the form

$$f(z) = ze^{-\alpha z} \qquad z \geq 0$$

$$= 0 \qquad z < 0$$

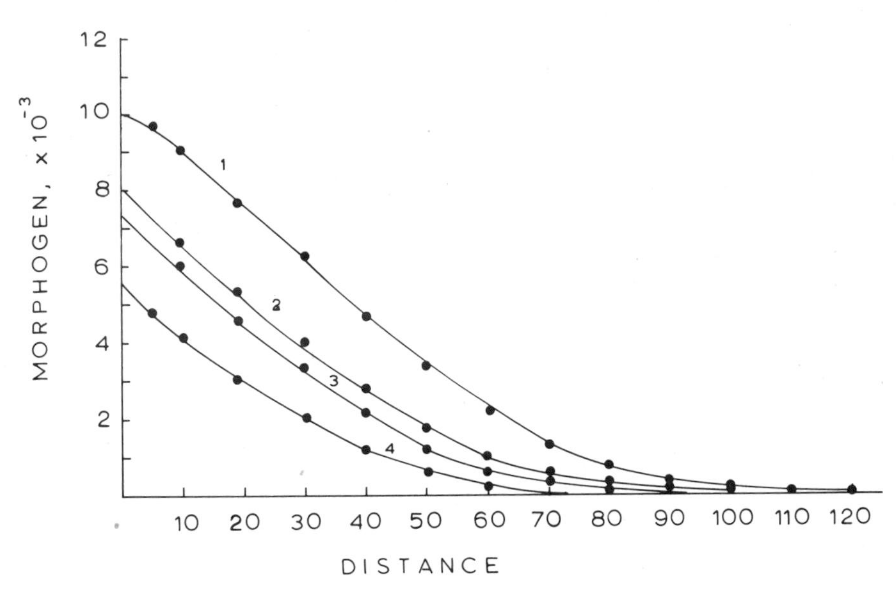

Figure A.1. Curves showing the spatial distribution of morphogen as computed from a particular form of the wave propagation model, described in the text, for different values of the delay, σ. The delay for curves (1), (2), (3) and (4) was 2·5, 4·0, 4·5 and 5·5 s respectively.

where $z = vt - v\sigma - x$ and α was chosen to give a maximum to the function 20 μm behind its origin. In these computations β was taken to be equal to zero. The different curves show the distribution of morphogen some time after wave initiation for different values of the delay, σ. As the delay increases, the amount of V picked up and converted into bound complex, the morphogen, decreases, since V is constantly destroyed. The curves all have stationary or near-stationary values of the morphogen at the origin, due to the property shown in Figure A.2 for the particular case of curve (3), when the delay is 4·5 s. One sees here that by 12 s after the initiation of the pick-up function, the value of the morphogen at the origin has reached about 96% of its asymptotic value. The same property holds for the other curves. It is this aspect of the model which makes it insensitive to axial lengths greater than some minimum, determined by the parameters, so that there is a length independent or regulative feature in its behaviour.

The essential properties of the wave model which generate the above results are easily described. An activity wave propagates along the membrane or cortex so that there is a moving source producing a diffusible metabolite at a fixed rate. This substance initially rises in concentration throughout the available space, diffusing everywhere, but then, due to the assumption that it is destroyed everywhere by the enzyme E_2 which is uniformly distributed in space, its concentration begins to fall and then goes

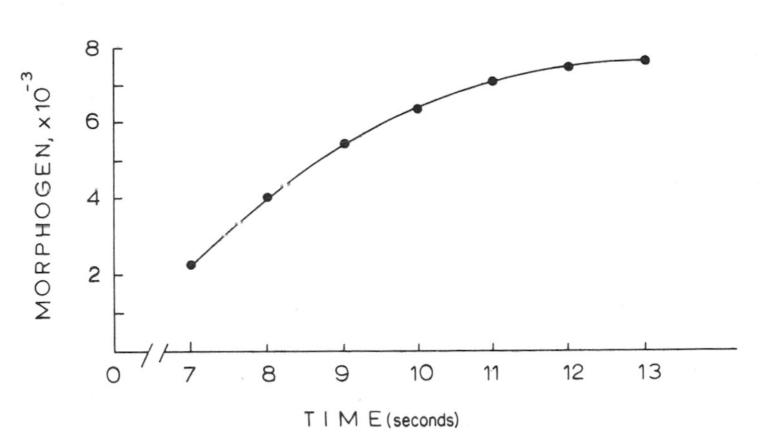

Figure A.2. The curve shows the rate of increase of morphogen at the origin as a function of time after initiation of the pick-up process.

on falling as its degradation rate exceeds its production rate. The pick-up process also propagates and converts $X(x, t)$ into $Y(x, t)$ in accordance with some kinetics, described in the model by the function $f(z)$, so that a spatial distribution of $Y(x, t)$ results. If there is a relatively long delay between the initiation of the wave and the commencement of the pick-up process, then $X(x, t)$ will be decaying relatively rapidly (greater sink than source, the former increasing as the wave propagates, the latter being constant), and a curve such as (4) results. However, if the delay is short, we get a curve such as (1).

There are, of course, many other ways of realizing the general principles of a wave model of the type described in Chapter 5, and some variants are under study. In general, the process seems to be basically simple and robust, and to have useful properties such as the regulative behaviour described above.*

* I am indebted to P. Cunningham for the computations reported in this appendix.

SUBJECT INDEX